The Natural Immune S

The Natural Killʋ ʋ

The Natural Immune System

The Natural Killer Cell

Edited by
CLAIRE E. LEWIS
and
JAMES O'D. McGEE

Nuffield Department of Pathology and Bacteriology
University of Oxford

OXFORD UNIVERSITY PRESS
Oxford New York Tokyo

Oxford University Press, Walton Street, Oxford OX2 6DP

Oxford   New York   Toronto
Delhi   Bombay   Calcutta   Madras   Karachi
Petaling Jaya   Singapore   Hong Kong   Tokyo
Nairobi   Dar es Salaam   Cape Town
Melbourne   Auckland

and associated companies in
Berlin   Ibadan

Oxford is a trade mark of Oxford University Press

Published in the United States
by Oxford University Press, New York

A catalogue record for this book is available from the British Library

Library of Congress Cataloging in Publication Data
The natural immune system : the natural killer cell / edited by C.
E. Lewis and J. O'D. McGee.
Includes bibliographical references and index.
1. Killer cells.   2. Natural immunity.   I. Lewis, Claire E.
II. McGee, James O'D.
[DNLM: 1. Killer Cells, Natural—physiology.   WH 200 N285]
QR185.8.K54N365   1992   616.07'9—dc20   91–32245

ISBN 0–19–963233–2 (h/b)
ISBN 0–19–963232–4 (p/b)

Set by
Footnote Graphics, Warminster, Wiltshire
Printed by
Information Press, Eynsham, Oxford

Foreword

Eva Lotzová, *Florence Maude Thomas Professor of Cancer Research, University of Texas, USA*

Although natural killer (NK) cells were seen as an immunological enigma for a number of years, the last decade has witnessed a notable reform in this perception. NK cells have now become established as a unique lymphocyte subset associated with a wide array of important biological functions. According to current definition, an NK cell is a large lymphocyte with densely staining azurophilic granules (i.e. a 'large granular lymphocyte', LGL), expressing CD16 and CD56 cell surface molecules and lacking CD3 and T-cell receptor ($\alpha\beta$ and $\gamma\delta$) structures. Although these are the typical features of most 'naive' (unstimulated) NK cells, variations in expression of surface phenotype as well as in morphological and functional NK cell characteristics have been observed. Such variations indicate the heterogeneity of NK cell populations. Whilst NK cells perpetually lack the CD3 antigen and T-cell receptor chains—features that distinguish them from T-cells—they may vary in their expression of other surface structures, including the 'notorious' NK cell molecules, CD16 and CD56. Both $CD16^+$, $CD56^-$ and $CD16^-$, $CD56^+$ NK cell variants have been detected in various tissue compartments.

In addition to variations in cell surface phenotype, NK cell populations are diverse both morphologically and functionally. Despite the strong association of NK cell activity with LGL, medium size and agranular NK cells have been identified. Furthermore, although NK cells have been described as nonadherent, this view now requires revision as recent studies have shown that a high proportion of NK cells adhere to plastic shortly after activation with IL-2 (termed 'adherent lymphokine activated killer cells', A-LAK). It is not clear whether such heterogeneity in NK cell characteristics reflects the existence of different NK cell subsets or different stages of NK cell differentiation/activation, although the latter appears more likely at present.

NK cells have been recognized primarily for their rapid oncolytic function, a cellular mechanism which is independent of the major histocompatibility complex (MHC) and any deliberate stimulation. Such oncolysis is directed against a variety of tumour targets and can be further enhanced and extended to include a wider spectrum of tumour cells after activation with various cytokines, the most potent of which is IL-2. Levels of NK cell oncolytic activity vary depending on the type of tumour involved. Some neoplastic cells have been defined as 'NK-resistant' (i.e. resistant to NK cell oncolysis), although this appears to be a rather relative phenomenon since resistant tumours can become susceptible after the enrichment or activation of the NK cell population. Moreover, the oncolytic activity of unstimulated NK cells, as measured *in vitro*, may be underestimated as NK cells could conceivably be more active *in vivo* under the continuous stimulation of natural, *in vivo*-produced IL-2, interferons, and/or other cytokines. This view is supported by data illustrating that NK cells can delay or prevent the *in vivo* spread of those metastatic tumours previously classified by *in vitro* criteria as NK-resistant.

Two important points need to be clarified in the context of this form of NK cell activity. First, despite the obvious phenotypic differences between T-cells and NK cells, IL-2-activated oncolytic T-cells are occasionally and incorrectly interchanged with NK cells. This comparison is based only on functional criteria, such as MHC-nonrestricted cytotoxicity, which encompasses lysis of NK-sensitive targets. Since more than one effector population may mediate oncolysis of the same tumour targets, such an analogy is unacceptable. T-cells with MHC-nonrestricted cytotoxicity should be clearly distinguished from NK cells. Second, misconceptions continue concerning the nature of the effector cell mediating the lymphokine-activated killer (LAK) cell phenomenon (the ability of IL-2-activated lymphocytes to destroy fresh tumour cells). LAK is still perceived by some as either a new, unique lymphocyte population and/or as a T-cell subset. Both of these views are inaccurate. It has been reproducibly shown that NK cells are the primary effector cells in the LAK phenomenon, whilst T-cells appear to play a rather minor role. Thus, it is more accurate to define LAK as an activity that can be assigned to the NK (NK-LAK) or T-(T-LAK) cell category, only after delineation of effector cells. Similarly, analysis of oncolytic IL-2-activated lymphocytes generated from solid tumours (tumour-infiltrating lymphocytes, TIL) has shown that TIL do not represent a new population of lymphocytes, but are composed of NK cells and T-cells. Which of these lymphocytes mediate oncolytic activity again appears to depend on the type of tumour.

It is of note that NK cells may be effective in the treatment of cancer, as suggested by their primary role in LAK activity. Conceivably, the adoptive transfer of purified IL-2 activated NK cells—rather than of the whole mononuclear cell population (where NK cells are in minority)—may result in a higher therapeutic benefit.

Soon after the discovery of NK cells, it became evident that anti-tumour activity was not their exclusive function. Various chapters in this volume present evidence on different NK cell activities. Both *in vivo* and *in vitro* studies demonstrate the involvement of NK cells in antimicrobial resistance; encompassing viruses, bacteria, fungi, and parasites. Considerable *in vivo* evidence of NK cell antiviral activity in humans exists and severe, often recurrent viral infections occur in individuals with selective NK cell deficiency. Other biologically relevant activities ascribed to NK cells are the regulation of hematopoietic and lymphoid cells, and the production of a variety of cytokines (e.g. IL-1, IL-2, IL-4, IFNs, TNF α and β, colony-stimulating factor, and macrophage-activating and chemotactic factor). These NK cell-derived cytokines have numerous widespread actions including direct anti-tumour effects, autochthonous NK cell regulation, as well as the regulation/activation of cells in the haematopoietic and lymphoid compartments. Less direct evidence suggests that NK cells also participate in the graft-versus-host reaction and in several autoimmune diseases.

NK cells comprise an important group of lymphocytes in human peripheral blood. However, they are also ubiquitously distributed amongst other body compartments, such as the lungs, liver, spleen, intestinal mucosa, and peritoneal cavity. Further, NK cells are similarly distributed in a number of animal species, and are involved in immune responses which phylogenetically precede B and T-cell responses. These findings indicate that NK cells may play a vital role in the homeostasis of the organism. Such a role may be accomplished by the aforementioned ability to regulate microbial infections and the growth and/or differentiation of haematopoietic and lymphoid cells. It is conceivable that involvement of NK cells in anti-tumour resistance may have evolved as a specialization of this earlier, primary function.

Whilst research of the last few years has broadened our knowledge of the phenotype and functions of NK cells, only some aspects of the cytolytic action of NK cells and its regulation have emerged. Little information is available concerning the target structures relevant for NK cell oncolysis, and the molecules and processes involved in NK cell recognition and transmembrane signalling. Clearly, NK cells mediate anti-tumour effects by more than one mechanism. The antibody and Fe receptor-dependent killing represents one cytotoxic pathway, and the putative NK cell receptor(s) may be involved in other pathway(s) of NK cell oncolysis. Moreover, anti-tumour effects can be mediated by NK cell-derived cytokines, some of which are known to be tumouricidal. Similarly, a number of diverse cellular mechanisms are likely to mediate other NK cell functions.

This brief introductory note merely touches on the immunobiological, molecular, and functional aspects of NK cells; the reader is invited to view individual chapters of this book for up-to-date reviews of current knowledge in these major topics of NK cell biology.

Acknowledgements

We would like to thank the following individuals for the support, advice and/or forbearance in the production of this book: Dr Louise Muntz, Mrs Anne McGee, Dr Frances Lannon, Dr Johann Lorenzen, and Dr Colette O'Sullivan.

Claire Lewis is also pleased to have the opportunity to thank Dr John Morris of the Department of Human Anatomy in Oxford. This book would not have been possible without his scientific and editorial guidance in the past.

We are also indebted to both Dr John Ortaldo for providing us with such an interesting illustration for the front cover, and to Professor Eva Lotzova for the erudite foreword which sets the scene for these collected works.

The idea for this book (as one of a series on the *Natural immune system*) was conceived during various stimulating discussions between the editors and representatives of the Scientific Publications Division of the Oxford University Press. We thank these individuals for this contribution, and are pleased to acknowledge their help and encouragement in the production of this volume.

The continuing support of the Cancer Research Campaign, UK, is also gratefully acknowledged by the editors.

Oxford
January 1992

C.E.L.
J.O'D.M.

Contents

3   Natural killer cells in bacterial infection

P. GARCIA-PEÑARRUBIA

4   Natural killer cells in viral infection

R. M. WELSH AND M. VARGAS-CORTES

5 Natural killer cells in transplantation

P. G. HOGAN

Contributors

P. Garcia-Peñarrubia Departmento de Bioquimica, Facultad de Medicina, Universidad de Murcia, 30100 Murcia, Spain

P. G. Hogan Lions Human Immunology Laboratories, Princess Alexandra Hospital, Ipswich, Wooloongabba, Brisbane, Queensland, Australia 4102

C. E. Lewis Nuffield Department of Pathology and Bacteriology, University of Oxford, John Radcliffe Hospital, Headington, Oxford OX2 9DU, UK

J. O'D. McGee Nuffield Department of Pathology and Bacteriology, University of Oxford, John Radcliffe Hospital, Headington, Oxford OX2 9DU, UK

J. R. Ortaldo Laboratory of Experimental Immunology, Biological Response Modifiers Program, Division of Cancer Treatment, National Cancer Institute, Frederick Cancer Research Facility, Frederick, Maryland 21702-1201, USA

J. O'Shea Laboratory of Experimental Immunology, Biological Response Modifiers Program, Division of Cancer Treatment, National Cancer Institute, Frederick Cancer Research Facility, Frederick, Maryland 21702-1201, USA

P. M. Starkey Nuffield Department of Obstetrics and Gynaecology, University of Oxford, John Radcliffe Hospital, Headington, Oxford OX3 9DU, UK

G. Trinchieri The Wistar Institute of Anatomy and Biology, 3601 Spruce Street, Philadelphia, Pennsylvania 19104, USA

M. Vargas-Cortes Department of Pathology, University of Massachusetts Medical Center, 55 Lake Avenue North, Worcester, Massachusetts, MA 01655, USA

R. M. Welsh Department of Pathology, University of Massachusetts Medical Center, 55 Lake Avenue North, Worcester, Massachusetts, MA 01655, USA

Abbreviations

ADCC	antibody-dependent cellular cytotoxicity
AIDS	acquired immunodeficiency syndrome
AKC	activated killer cytotoxicity
AR	allogeneic resistance
ARC	AIDS-related complex
ATP	adenosine triphosphate
BCG	bacille Calmette–Guérin
BFU	burst-forming unit
BLT	butoxycarbonyl-lysil-tyrosine-esterase
BMG	bone marrow graft rejection
BPA	burst-promoting activity
CD	cluster of differentiation
CFU	colony-forming unit
CMV	choriomeningitis virus
CRF	corticotrophin-releasing factor
CSF	colony-stimulating factor
CTL	cytotoxic lymphocyte
DC	dendritic cell
EBV	Epstein–Barr virus
EGF	epidermal growth factor
EGFR	epidermal growth factor receptor
EPO	erythropoietin
FAA	flavone acetic acid
FACS	fluorescence-activated cell sorting
GALT	gut-associated lymphoid tissues
GAP	guanine nucleotide-activating protein
GM	granulocyte-macrophage
GM-CSF	granulocyte-macrophage colony-stimulating factor
GVHD	graft-versus-host disease
H	histocompatibility
Hh	haematopoietic histocompatibility
HIV	human immunodeficiency virus
HLA	human leucocyte antigen
HR	hybrid resistance

ICAM	intercellular adhesion molecule
IFN	interferon
Ig	immunoglobulin
LAD	leucocyte adhesion deficiency
LAK	lymphokine-activated killer
LCMV	lymphocyte choriomeningitis virus
LPS	lipopolysaccharide
MAB	monoclonal antibody
MAF	macrophage-activating factor
MCMV	murine cytomegalovirus
MDC	monocyte-depleted cell
M-CSF	macrophage colony-stimulating factor
MDP	muramyl dipeptide
mH	minor histocompatibility
MHC	major histocompatibility complex
MLC	mixed lymphocyte culture
MLR	mixed lymphocyte reaction
MTP-PE	muramyl tripeptide-phosphatydyl ethanolamine
NA	neuroaminidase
NCAM	neural cell adhesion molecule
NK	natural killer (cell)
NKCF	natural killer cytotoxic factor
NKR	natural killer (cell) receptor
PBL	peripheral blood lymphocyte
PBMC	peripheral blood mononuclear cell
PDGF	platelet-derived growth factor
PDGFR	platelet-derived growth factor receptor
PGM	peptidoglycan monomer
PHA	phytohaemagglutinin
PI	phosphatidylinositol
PKC	protein kinase C
PLC	phospholipase C
PPD	purified protein derivative
PTK	protein tyrosine kinase
scid	severe combined immunodeficiency
SEB	staphylococcal enterotoxin B
SEM	scanning electron micrograph
TCR	T-cell receptor
TEM	transmission electron micrograph
TGF	transforming growth factor
TIL	tumour-infiltrating lymphocyte

TNF	tumour necrosis factor
TR	transferrin receptor
VDCC	virus-dependent cell-mediated cytotoxicity
VIP	vasoactive intestinal peptide
VSV	vesicular stomatitis virus

1 The biology of natural killer cells: insights into the molecular basis of function

J. O'SHEA and J. R. ORTALDO

1   Introduction

Studies regarding natural cytotoxic activity had their origin in the early 1970s when spontaneous cytotoxic activity was observed in leukaemic twins (1). In these studies, normal immunized twins demonstrated background lytic activity during the search for a specific cell-mediated antitumour response. These studies led to numerous in-depth studies of human effector cell systems (2–4) that were the basis for the discovery of endogenous spontaneous cytotoxicity, later termed 'natural killer' or 'NK' activity (5). From these beginnings, a great deal of work regarding the study of NK activity has accumulated over the last decade. Subsequent studies have identified the cell type responsible for this activity as being a large granular lymphocyte (LGL). This form of lymphocyte has a high cytoplasm to nuclear ratio and displays distinct azurophilic granules (6). More recent studies have outlined both the phenotype (7–9) and the functional regulation of LGLs [reviewed in (10–12)]. However, it should be noted that, as with other cell types, heterogenicity exists amongst the NK cell population with some NK cells not displaying the distinct LGL morphology during certain states of differentiation and/or activation [for references, see (11)].

With the wide array of studies performed over the last decade on spontaneous cell-mediated cytotoxicity, it has become clear that natural effectors comprise a variety of cell types which mediate both distinct cytolytic capacities as well as non-cytolytic functions. These capacities include not only NK activity, but also lymphokine-activated killer (LAK) activity and antibody-dependent cellular cytotoxicity (ADCC). Spontaneous non-major histocompatibility complex (MHC) restricted killing can be mediated by natural killer (NK) cells, certain activated T-cells, and cells in the monocyte/macrophage series.

The focus of this review will be the regulation of NK cell function and, where possible, the molecular basis of various functions will also be described. In addition, correlations will be made between the function of NK cells and that of other cytolytic effector cell types. Since the description of the NK cell as being a LGL, some of the most exciting recent findings in this area have involved the detailed biochemical analysis of membrane molecules that are critical to NK cell function. Such studies have extended previous investigations performed in the early 1980s into the role of interferons and interleukins in the regulation of NK cells, to begin examining signal transducing molecules. These have included both the characterization of the receptors for interleukin-2 (IL-2) and the involvement of various IL-2-stimulated signal transducing agents in the activation and proliferation of NK cells. Another exciting development has been the refinement in the understanding of the structure and function of the NK Fc receptor and its association with the ζ chain (13–15). It is encouraging to see the progression in the study of NK cells from the level of cellular phenomenology to a truly molecular arena. With these numerous ongoing studies regarding NK cells, it is obvious that receptors involved not only in regulation of function, but also in recognition and cytotoxicity will soon be enumerated.

2 Definition

The best definition of NK cells has emerged from a recent, 1989 Natural Killer Cell Workshop held in Hilton Head, (North Carolina, USA) that attempted to define and place NK cells among other cell functions and lineages. This definition was based on a consensus approval by researchers studying NK cells (16).

NK cells are $CD3^-$ T cell receptor (alpha, beta, gamma, delta)$^-$ large granular lymphocytes. They commonly express certain cell surface markers such as CD16 and NKH-1 (Leu-19) in humans and NK-1.1/NK-2.1 in mice. They mediate cytolytic reactions that do not require expression of class I or class II MHC molecules on the target cells.

Certain T lymphocytes which are either alpha/beta$^+$ or gamma/delta$^+$ may express, particularly upon activation, a cytolytic activity that resembles that of NK cells. These T lymphocytes should not be termed NK cells. They could be termed either T lymphocytes displaying "NK-like" activity or "non-MHC requiring" cytolysis.

Lymphokine-activated killer (LAK) cells as IL-2-activated lymphocytes in either of the two above categories. The relative contribution of the respective cell type depends on the source of lymphocytes and conditions for activation. For instance, lymphocytes from peripheral blood or spleen will produce lymphokine-activated killer cells from NK cells in a close to predominant manner.

From this definition it is obvious that cells other than NK cells can mediate non-MHC restricted killing against target cells termed NK susceptible. The predominant cell type mediating non-MHC restricted lysis, however, is the $CD3^-$ LGL. Although analogous cells exist in rodents and other animal species, this review will focus mainly on the function of NK cells in the human system.

3 Characteristics

The general characteristics of the cytotoxicity displayed by lymphocytes mediating natural effector activity are summarized in Table 1.1. This lytic activity can be observed against a variety of neoplastic and viral-infected cells in a non-MHC restricted fashion. In addition, NK cells have been shown to be capable of ADCC. Although effector cells can mediate non-MHC restricted lysis, even in the absence of activation, they can also be

Table 1.1 General characteristics of the cytotoxic functions of natural effector cells

Characteristic	NK	ADCC	LAK	
			CD3$^-$	CD3$^+$
1. Spontaneous	Yes	Yes	No	No
2. Major effector cell				
(a) morphology	LGL	LGL/Mφ/PMN	LGL	LGL, usually CD8$^+$
(b) receptor	?	FcγR	?	TCR?, other receptors
3. Target types	Some tumours, virus-infected, stem cells	Ab-coated cells	All tumours	All tumours
4. MHC-restricted	No	No	No	No
5. Requires Ab	No	Yes	No	No
6. Augmenting agents	IFNα,β,γ IL-2	IFNα,β,γ IL-2	IL-2	IL-2

?, unknown or undefined; ADCC, antibody-dependent cellular cytotoxicity; LAK, lymphokine-activated killing.

stimulated with IL-2 to display potent NK and LAK forms of cytotoxicity. This type of killing is usually mediated by CD8⁺ T-cells. However, the receptors utilized by these effector cells have yet to be clearly defined.

The first and most direct study to implicate a defined cell population to NK activity involved the observation that sedimentation of LGLs in discontinuous Percoll gradients coincided with cells having cytolytic activity in this gradient (6). Although minor differences in the properties of LGLs exist between species, certain features indicate that this unique cell type is clearly responsible for most of the NK activity in not only the human but in the mouse and the rat systems (17–20). Most of the characteristics of NK cells for these three species are summarized in Table 1.2 (11). In general, in all species studies, LGLs are non-adherent cells that possess Fc receptors but lack immunoglobulin- and T-cell receptor-gene rearrangement.

Table 1.2 Phenotypic characteristics of NK cells

	Characteristics
1. General	
Size	Large, 12–15 μm (mouse, 8–10 μm)
Cytoplasmic granules	+
Kidney-shaped nucleus	+
Adherence	−
Phagocytosis	−
2. Histochemistry	
Acid phosphate	+
Non-specific esterase	+
β-Glucuronidase	+
Peroxidase	−
TdT	−
3. Cell surface antigen	
CD2	≥75% (rat 30%)
CD3	≤5%
CD4	≤1%
CD5	≤5%
CD7	80% (NT in rodents)
CD8	25% (<1% in mouse, >90% in rat)
CD11	>90% (NT in rodents)
CD16	50–90% (ADCC functionally weak in rodent) [FcγRIII MAB not available in rodent]
FcμR	<1%
sIg	<1%
Ia	10% (≤1% in mouse)
Asialo GM1	>90%

Discordance between species indicated by parentheses. +, >90% of the cells are positive for this characteristic; −, negative for characteristics; NT, not tested.

Morphologically, these cells are characterized as LGLs with a kidney-shaped nucleus with prominent azurophilic granules. Murine NK cells are smaller than human LGLs, but nonetheless have a granular lymphocyte morphology. A variety of laboratories have extensively characterized the surface phenotypes of NK cells (7–9) and demonstrated that human LGLs share both myelomonocytic- (e.g. CD11) and T-cell- (e.g. CD2 and CD8) related markers. However, the majority of human NK cell activity is mediated by $CD3^-$, $CD16^+$, $CD56^+$ lymphocytes. Due to their expression of CD16 (immunoglobulin Fc receptor, $Fc\gamma RIII$) these cells exhibit high levels of ADCC activity. However, $CD16^-$ NK cells also exist that have LGL morphology and express markers similar to $CD16^+$ NK cells, including: CD2, CD7, CD11b, CD38, CD45R, CD18, and p75 IL-2R (8–21). The cells are active cytolytically and respond to IL-2 in terms of proliferation and activation. In fact, $CD16^-$ NK cells preferentially proliferate in response to IL-2. However, cytokine gene expression is lower in $CD16^-$ NK cells. This is particularly true with respect to their production of interferon γ ($IFN\gamma$) (21). Most NK cells in the mouse do not express the CD4 and CD8 antigens on their cell surface but do express selective markers, such as asialo-GM_1, NK1.1, and NK2.1 or LGL-1. Little is known of the expression of CD56 by NK cells in mice, as monoclonal antibodies (MABs) directed against the murine homologue of CD56 have not yet been derived.

The lytic function receiving most attention recently is the ability of NK cells to be activated by IL-2 to mediate LAK activity. In contrast to resting NK cells, effector cells mediating LAK activity are able to kill virtually all tumour cells and virus-infected cells. Such cells have little or no effector activity against non-malignant or uninfected normal counterparts. Studies have demonstrated that the phenotype of cells mediating LAK activity in the human, rat, and mouse appears to be synonymous with the NK cell, as $CD3^-$ lymphocytes. Analysis of effector cells isolated by progenitor analysis, cell sorting, gradient separation, and limiting dilution analysis have been consistent with the view that $CD3^-$ LGLs mediate the majority of lymphokine-activated killing (22–24). Therefore, IL-2 expansion and conversion of these cells into highly effective lymphokine-activated killer cells have been areas of intense study. The ability of IL-2 to regulate NK cell function has been clarified by recent studies using monoclonal antibodies to the beta chain of the IL-2 receptor ($IL-2R\beta$) (25, 26). These monoclonals have verified previous studies regarding IL-2 activation of NK cells and have solved the enigma of IL-2 activation of cells in the absence of TAC ($IL-2R\alpha$) expression. It is now clear that NK cells spontaneously express high levels of $IL-2R\beta$ and thereby exist as a unique population of effector cells with the ability to respond rapidly to IL-2 and IL-2-induced signals.

In addition to the cytolytic functions that have been most thoroughly studied over the last 15 years, evidence now exists that $CD3^-$ LGLs exhibit

Table 1.3   Functions of NK cells

1. Control of tumour cell growth
2. Involvement in the control of microbial infections
   (a) Viral infections
   (b) Parasites (intracellular and extracellular)
   (c) Fungi
   (d) Bacteria
3. Immunoregulatory properties
   (a) Production of cytokines
   (b) Control of hematopoietic stem cell growth and differentiation
   (c) Involvement in allograft rejection
4. Disease states
   (a) Involvement in the development of graft vs. host (GvH) disease
   (b) Contribute to some forms of aplastic anaemia/neutropenia
   (c) Potentiate autoimmune and neurological disease
   (d) Contribute to the development of some forms of diabetes
   (e) Involved in various gastrointestinal disease

key non-cytolytic functions as well (summarized in Table 1.3) (see later chapters in this volume). These functions include a number of important immunoregulatory properties. The best studied non-cytolytic function of CD3⁻ LGLs is the ability of these natural effectors to control the spread of microbes, especially viral particles. Available data strongly support the conclusion that NK cells play a critical role in eliminating the initial replication of various forms of viruses including human cytomegalovirus, herpes simplex type I, and murine hepatitis virus (11, 12). Also, NK cells have been shown to have direct and indirect effects on some bacteria (27, 28) as well as extracellular and intracellular parasites (29). These findings suggest an important role for this cell type in limiting the growth and spread of a variety of microbial infections. Additionally, these observations may indicate that LGLs and their products account for a significant proportion of the host's inflammatory response in their role in the irradication of microbes.

Studies carried out in the 1970s by Cudkowicz and colleagues also indicated a role for NK cells in the regulation of haematopoiesis (30). The observations that immature cells from the bone marrow or thymus were good targets for NK-mediated cytolysis provided support for this hypothesis (11, 30–32). More recent studies have shown that lymphocytes with NK activity are able to inhibit the development of bone marrow stem cells *in vitro* and *in vivo*. In many cases the level of NK activity correlated with the ability of bone marrow to reconstitute irradiated recipients (33). Although the basic mechanisms for these observations regarding the regulation of the haematopoiesis have not been clearly defined, one possibility resides in the ability of LGLs to produce a variety of cytokines. CD3⁻

LGLs have been shown to secrete interleukin-1 (IL-1), IL-2, B cell growth factor (or interleukin-4; IL-4), interferon α, tumour necrosis factor (TNK) α and β, various colony stimulating factors, and a macrophage-activating and chemotactic factor (10, 11, 33). The production of these cytokines suggests that LGLs play a more general regulatory and/or developmental role in the many humoral or cellular responses mediated by T-cells, B-cells, and other non-lymphocyte populations. In addition to their cytolytic capabilities, the demonstration that LGLs produce these immunoregulatory molecules raises the possibility that this cell type may play an important *in vivo* role in controlling immune and inflammatory responses.

4   Origin, differentiation, and distribution

Various experimental systems, both *in vivo* and *in vitro*, have provided clues to the cell type(s) that originate and differentiate into inactive NK cells from precursor bone marrow cells. This was shown to be an inherent, autonomous property of bone marrow in early murine studies involving the manipulation of mutant strains of mice with congenital abnormalities in NK cell function, or bone marrow reconstitution of radiation chimeras. Although a detailed immunophenotypic analysis of such bone marrow precursor NK cells has yet to be performed, a number of recent studies have shown that IL-2 stimulates the proliferation of $CD2^+$ (partly $CD16^+$) bone marrow cells with NK cell activity.

Transfer experiments in mice have suggested that NK cells further differentiate in the peripheral blood, finally reaching a large non-proliferating, resting state either in the blood or in residence in one of a variety of body tissues. Although considerable inter-individual variation in the NK cell content of human peripheral blood exists, on average, 9–10 per cent of peripheral blood lymphocytes are $CD16^+$, $CD56^+$ cells with LGL morphology and NK activity. The spleen (red pulp) is quite rich in NK cells, which may constitute up to 5 per cent of the total lymphocyte population of the spleen. Relatively, fewer mature, cytotoxic NK cells are present in white pulp of the spleen, thymus, lymph nodes, tonsils, and bone marrow. Cells with NK activity can be detected in most tissues (at least in long-term *in vitro* cytotoxicity assays) and are particularly found in the lung interstitium, intestinal mucosa, peritoneal exudate, and liver sinusoids. However, owing to the marked heterogenicity of immunophenotype which can exist among NK cells, a panel of antibodies that recognize CD2, CD8, CD16, CD56, and CD57 (all of which can be expressed by NK cells) are being used in current studies to identify NK cells in such tissues unambiguously.

Adoptive cell transfer studies using radiolabelled rodent NK cells have shown that NK cells do not recirculate in the thoracic lymph and are relatively short-lived cells. Any increase in NK cell activity within a given tissue following immunostimulation reflects either the proliferation or enhanced activity of resident NK cells, infiltration of NK cells from the peripheral blood or bone marrow precursors, or redistribution of NK cells from another tissue (for a more extensive coverage of this topic, the reader is referred to the recent review by Trinchieri (11)).

5 Biochemical basis of NK cell activation

The last and major part of this chapter will focus on the subcellular mechanisms subserving the activation of resting mature NK cells in the general circulation or body tissues. Activation may involve one or a number of cellular activities, such as proliferation, cytotoxicity and/or cytokine secretion.

The process by which NK cells become activated to lyse their targets and secrete cytokines is comprised of several distinct steps. First, NK cells must recognize targets presumably by receptor molecules on their cell surfaces. Secondly, these molecules, in turn, must transmit a signal across the cell membrane indicating that they have been ligated. Thirdly, this signal must activate the cytolytic and secretory machinery and/or result in the alteration of gene expression. In addition, cytokines, by virtue of interacting with their respective receptors, can modulate the function of NK cells by transmembrane signalling and the concomitant alteration of gene expression. NK cell activation then is comprised of multiple biochemical steps and may involve a panoply of receptors (Figure 1.1). Because of the inherent ambiguity, investigators need to be wary of the term 'NK cell activation'. Clearly, the recognition of tumour targets by virgin NK cells results in an activation programme which results in the lysis of targets. Treatment of the same cells with IL-2 also activates them. However, surface molecules, as well as the biochemical intermediates responsible for these discrete forms of so-called activation, are not the same. Additionally, the term 'activation' does not specify a particular event; it may refer to the generation of second messengers, alteration of gene expression or changes in cytolytic capacity. The precise measure of activation is not inherently apparent. However, despite its potential for ambiguity and confusion, the term activation is convenient, providing one attempt to clarify the receptor that is engaged and the measure of activation utilized. An additional caveat in the field of NK activation is that the requirements for activating freshly isolated NK cells may differ substantially from those of either NK clones grown *in vitro* or NK-like cell lines. Thus the cell source also needs to be kept in mind in studies of NK cell activation.

The biochemical basis by which engagement of the T-cell receptor (TCR) functions in transmembrane signalling has received considerable

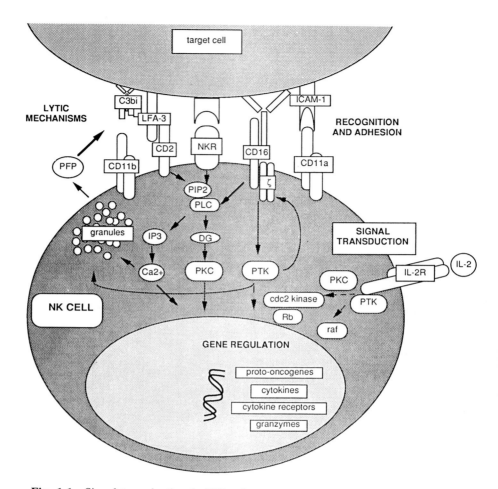

Fig. 1.1 Signal transduction in NK cells. Both the documented and speculative signalling events employed by CD3⁻ LGLs are depicted. Abbreviations include: DG, diacylglycerol; IP_3, inositol trisphosphate; NKR, putative NK receptor; PFP, pore-forming protein; PIP_2, phosphatidylinositol bisphosphate; Rb, retinoblastoma gene product.

attention (34–36). Although the structure and function of this receptor are likely to be quite different from those molecules utilized by NK cells to recognize and react to target cells, recent data have indicated that the two cells share surprising similarities. In addition, since T-cells mediate MHC-restricted, as well as non-MHC-restricted, cytolysis the majority of data pertaining to signalling in the CTL has relevance to NK cells. However, NK cells also have a wide range of unique properties which are reflected, in part, by distinct modes of signalling. Such similarities and differences in the molecular events involved in signalling in T- and NK cells are discussed in detail below.

5.1 Surface structures involved in triggering NK cells

NK cells recognize and lyse a limited number of tumour cell types. The principal structure(s) involved in this process remain(s) enigmatic. However, the structure of a number of other cell surface molecules on LGLs have been clearly defined. In contrast to the putative NK receptor, the structure and function of the receptor for the Fc portion of IgG on NK cells has recently emerged as a result of an outpouring of exciting new data. A second group of accessory and adhesion molecules on LGLs has also been quite well defined. The binding of T-cells to cells expressing antigen is mediated by a number of molecules in addition to the TCR, denoted as accessory molecules. Perturbation of many of these structures activates the T-cell. Analogously, ligation of several molecules on the surface of NK cells, generally with antibodies, also may be an activating stimulus for NK cells (Table 1.4). The predominant function of many of these molecules is to enhance adhesion, though it has become increasingly clear that so-called

Table 1.4 Signalling molecules on NK cells

CD nomenclature	Synonyms	Ligand	Mol. wt. ($\times 10^3$)	Signalling PI-Ca^{2+}	PTK
CD16	FcγRIII	IgG	55–70	++	+
CD2	SRBC receptor	LFA3	50–55	++	?
CD11a/CD18	LFA1	ICAM	α 150 β 95		
CD11b/CD18	C3biR/CR3	iC3b	α 180 β 95		
CD11c/CD18	p150, 95	?	α 150, β 95		
CD8	–	MHCI	34 kDa		
CD56	NKH1 Leu19	?	~220		
CD45	Leucocyte Common antigen	?	~200		
	LFA3	CD2			
	ICAM1	LFA1	90 kDa		
	35 kDa	?	35 kDa	++	
	42 kDa	?	42 kDa	++	
	3.2.3				
	GLA 138	?			
	IL-2Rβ	IL-2	p75		+
CD25	IL-2Rα	IL-2	p55		
—	Putative NK receptor	?	?	++	?

?, unknown or undefined.

adhesion molecules are not distinct from signalling molecules (1). Because of the intensity of recent research efforts, the NK Fc receptor will be discussed first.

5.1.1 The receptor for the Fc portion of immunoglobulin (CD16) on NK cells: the zeta and/or gamma connection

The ability of NK cells to lyse immunoglobulin- (Ig) coated targets is due to the expression of a receptor for the constant (Fc) portion of Ig on NK cells (FcR). Although it is beyond the scope of the present chapter to discuss in detail this heterogeneous group of receptors (37–39), these can be grouped into three types of FcR. FcγRI is a high affinity receptor expressed on monocytes, whereas FcγRII is a low affinity receptor expressed on B-cells, neutrophils, macrophages, and platelets. FCγRIII (CD16) is a low affinity receptor expressed on macrophages, neutrophils, eosinophils, and a subset of T-cells and NK cells. Using monoclonal antibodies (MABs) raised against anti-FcγRIII, this FcR was cloned by expression in COS (SV-40 transformed African green monkey kidney cell) cells. The amino acid sequence derived from cDNA analysis revealed it to be a phosphatidylinositol glycan-linked molecule, and transfection of this molecule conferred the ability to bind Ig (40). However, the molecular form of CD16 can differ amongst various cell types; specifically, the form in NK cells is slightly larger, a difference that is not accounted for by differences in glycosylation (41). Additionally, whilst the neutrophil form of the receptor is susceptible to cleavage by phosphatidylinositol-specific phospholipase C, the NK cell form is not (42–45). The issue of heterogeneity of CD16 was resolved by the demonstration that this type of FcR exists as two isoforms, a phospholipid-linked form and a transmembrane form, encoded by separate genes (46–48) that are differentially expressed in neutrophils and NK cells, respectively. The gene encoding the phospholipid-linked form in neutrophils differs from that encoding the transmembrane form of the molecule in NK cells by nine nucleotide substitutions, resulting in six amino acid differences and the absence of 21 amino acids at the intracellular carboxyl terminus. Of note, a single critical residue (Ser203) determines phospholipid linkage of FcRIII. The Fc receptor of rat NK cells has also recently been cloned and shares considerable homology with mouse and human Fc receptors (49). Interestingly, transfection of the cDNA for the transmembrane form of FcRIII failed to result in surface expression of the molecule. The explanation for these data is that additional subunits are required for efficient cell surface expression of the transmembrane form of FcγRIII. This discovery was preceded by the following observations. The Fc receptor for IgE, like the TCR, is a multi-subunit receptor and multiple subunits are required for optimal plasma membrane expression of the receptor complex (50). Of particular interest, one chain of the Fcε receptor shares homology with a chain of TCR. These chains are the gamma and zeta chains, respectively. Although it had previously been believed that the zeta chain was exclusively associated with the TCR, it has recently been shown that this so-called subunit of the TCR is expressed in NK cells (13). Subsequently, it was shown that at least one structure that zeta associates with in LGLs is FcγRIII (14, 51). When COS cells were transfected with the cDNA for both the zeta chain and FcγRIII, this then allowed surface expression of the latter. In a manner similar to zeta, it then became evident that the expression of the gamma subunit of FcεR is not restricted to cells bearing IgE receptor, such as mast cells. The gamma subunit is also present in mouse macrophages associated with FcγRIIa (52). In transfection experiments, surface expression of FcγRIIa required co-transfection with the gamma chain. (The nomenclature here is unfortunately confusing. As a subunit of the Fcε receptor, the gamma chain does not sound so cumbersome. However, since the discovery that the gamma chain is also a subunit of Fcγ receptors, the term is, in retrospect, an unfortunate designation.) Similarly, surface expression of the NK cell FcR, FcγRIII is also permitted if cells are co-transfected with the gamma chain (48, 53).

A further surprise (13) was that a new species of zeta is evident in LGLs. In the T-cell the majority of zeta exists as a homodimer. In murine T hybridoma cells a minority of zeta (about 10 per cent) exists as a heterodimer associated with eta, and this may have important implication in signal transduction (54). In LGLs a different heterodimeric form of zeta is present, zeta-p12 (Figure 1.2). Although the identity of this latter molecule has not been elucidated with any certainty, this may represent zeta disulphide linked to the gamma subunit of Fc receptors. This speculation is supported by the observation that such a species exists in T-cells (55). It has yet to be shown definitively, however, if the gamma chain associates physiologically with the FcγRIII of NK cells.

Thus it appears that several members of the Fc receptor family (Fcε, FcγRIIa, and FcγRIII), as well as the TCR, may have similar rules for receptor assembly. In addition, optimal transport and expression of plasma membrane receptors require more than a single subunit, and one subunit in particular (zeta, gamma, or the combination of the two) appears to be essential. As will be discussed below, zeta has critical roles not only in assembly of multimeric receptors but also in signal transduction. In addition to FcγRIII, it appears that murine LGLs also express FcγRII. This was demonstrated by showing that binding of the anti-FcγRII MAB, 2.4G2, to mouse NK cells inhibited the binding of immune complexes. In addition, mRNA for FcγRIIa was also detected in these cells (56).

5.1.2 Adhesion and accessory molecules on NK cells

The CD2 molecule is the molecule that enables T-cells to bind to sheep red blood cells and thus was the earliest marker of T-cells. Although this molecule is also expressed by LGLs, it is not found on cells of other lineages (57).

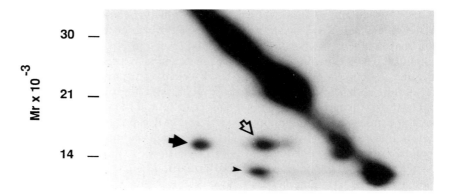

Fig. 1.2 Two-dimensional, diagonal gel analysis of surface expressed zeta (ζ) chain on NK cells. Isolated CD3⁻ LGLs were surface labelled with Na^{125}I, lysed, and immunoprecipitated with an anti-zeta serum. The immunoprecipitated proteins were eluted under non-reducing conditions and run on SDS-polyacrylamide tube gels. Tubes were equilibrated in a reducing agent and were then adhered to slab gels and run in the second dimension. Proteins migrating on the diagonal are not disulphide linked to other proteins. The proteins on the diagonal visualized in this gel on the diagonal are non-specifically immunoprecipitated. In contrast, since ζ is a disulphide-linked protein, it migrates below the diagonal on non-reducing/reducing gels. The species of zeta denoted by the large solid arrow is ζ–ζ homodimer and migrates with an Mr of 32 000 under non-reduced conditions, whereas following reduction, monomers run with a Mr of 16 000. The species of ζ denoted by the large open arrow is ζ which is disulphide linked to a protein with a Mr of ~10–12 000 (small arrow). The identity of this latter protein has not been clearly elucidated but may be the gamma chain. (I. C. S. Kennedy and co-workers, unpublished observations.)

This 50–58 kDa glycoprotein functions as an adhesion molecule and is a member of the Ig gene superfamily. The murine homologue of human CD2 has recently been cloned (58). As transfection of cells with CD2 alone results in surface expression of this molecule, it is unlike the TCR and FcR in the requirement for multiple subunits for expression. The ligand for this receptor is another adhesion molecule LFA3, a 60–70 kDa molecule that is also expressed on NK cells as well as a variety of other cell types. LFA3 exists as a glycan-linked and a transmembrane form.

In addition to its function in adhesion, CD2 appears to have signal transducing functions both in T-cells and in NK cells. Interestingly, most data in T-cells point to a requirement for expression of the TCR for efficient signalling via CD2 (57). However, this obviously cannot be the case for NK cells, and no solution to this enigma is readily apparent at this time. Several types of studies provide evidence for the signalling function of CD2 in NK cells (discussed in detail below) including redirected lysis assays (59). LFA3 has been shown to be a signalling molecule for mono-cytes inasmuch as perturbation of LFA3 triggers release of IL-1 and TNFα (60), but no data exist for such a role in NK cells.

LFA1 (CD11a), an integrin gene superfamily member, is also a member of a family of heterodimeric molecules (leucocyte adhesion proteins) that shares common beta chain (CD18) [reviewed in (61)], and is expressed on T-cells and NK cells. The ligand for LFA1, ICAM-1, is a 90 kDα transmembrane protein and is a member of the immunoglobulin gene superfamily. ICAM-1 has homology with the neural cell adhesion molecule (NCAM), an isoform of the NK cell protein, CD56. ICAM-1 is expressed on leucocytes, epithelial cells, endothelial cells, and fibroblasts. ICAM-1 expression is regulated by the cytokines, IL-1, IFNγ, and TNFα. Antibodies against LFA1 can inhibit NK cell-mediated killing although the effect is not complete (62). Antibodies against the β chain of leucocyte adhesion proteins, however, effectively inhibit cytoxicity (63), suggesting that other members of this family may contribute to the binding of NK cells to targets. The avidity of interaction between ICAM-1 and LFA-1 is not simply a passive process; perturbation of the TCR or CD2 and pharmacological activation of T-cells with phorbol esters results in enhanced LFA1 adhesion (64–66) and this interaction is likely to be analogously regulated in CD3⁻ LGLs.

The receptor for a breakdown product of the third complement com-ponent, C3bi, is expressed on neutrophils, monocytes, macrophages, and NK cells. This receptor is another member of the leucocyte adhesion protein family and comprises an alpha chain with homology to CD11a (LFA1) and CD11c (p150) and a beta chain (CD18) that is common to all of the members of this family. This receptor could potentially function in a similar manner to the NK Fc receptor in that particles opsonized by fragments of C3 may also be lysed by NK cells in a manner analogous to ADCC. Indeed, serum-coated targets appear to be more efficiently lysed by NK cells, an effect which can be blocked by anti-CD11b antibodies (67–69). In neutrophils and monocytes, CR3 expression is up-regulated by chemotactic peptides and activators of protein kinase C (PKC) (70), but this has not been shown in NK cells.

CD56 is often used as a definitive marker for NK cells. However, recently it has become clear that CD56 is an isoform of neural cell adhesion molecule (NCAM). The latter is a membrane glycoprotein ex-pressed on neural and muscle tissues involved in adhesive interactions in these tissues and is a member of the IG gene superfamily (71). Anti-CD56 antibodies have synergistic inhibitory effects with anti-LFA1 or anti-LFA3 on NK cell-mediated binding and cytotoxicity, and thus a variety of ad-hesion molecules appear to regulate the interaction of NK cells with their targets. NK cells also express the well-characterized adhesion proteins fibronectin and laminin. Antibodies against these proteins also inhibit NK cell-mediated cytotoxicity (72, 73).

Another molecule, also a member of the Ig gene superfamily, that is exclusively expressed on T-cells and NK cells is the CD8 molecule. In T-cells, this molecule is a marker for Class I MHC-restricted antigen recognition. CD8 interacts with Class I MHC proteins to enhance the avidity of the interaction between the T-cell and its target. Since the killing mechanism of NK cells is not dependent on expression of MHC molecules, the role of CD8 is unclear. In addition, only a subpopulation of LGLs bears the CD8 antigen and the membrane density expression of CD8 is low compared to that of $CD3^+$ T-cells. Interestingly, $CD8^+$ NK cells have reduced NK activity relative to $CD8^-$ NK cells (74). However, anti-CD8 antibodies neither triggered nor were involved in lytic function (75). Recently, it has become clear that CD4 and CD8 molecules are physically associated with a *src* family cytoplasmic tyrosine kinase, $pp56^{lck}$. Although this protein kinase is expressed in LGLs (H. Young, unpublished observations), the function of CD8 and $pp56^{lck}$ in LGLs is unclear at present.

The CD45 family of membrane proteins, a heterogeneous group of molecules found on hematopoietic cells, has received considerable attention recently. This is because the CD45 gene encodes a collection of transmembrane protein tyrosine phosphatases (77–80). The reason for the heterogeneity lies in multiple splice variants with cell-specific expression (81). The extracellular portions are altered among these splice variants suggesting receptor-like molecules. However, the ligands for this family of proteins have not been fully elucidated. The intracellular portion is conserved among the family members and contains a duplicated tyrosine phosphatase domain with homology to the placental protein tyrosine phosphatase. An extensive body of literature has emerged on the function of CD45 in T-cells. Signalling via the TCR, CD4, and CD2 molecules appear to be regulated by CD45 (82–84). In addition, physical association between CD2 and CD45, as well as TCR-associated zeta chain and CD45, has been demonstrated by cross-linking studies (84, 85). Moreover, CD45 appears to regulate the enzymatic activity $pp56^{lck}$ in T-cells (86–89). To date, there is little information on the function of CD45 in NK cells, but this is clearly a fertile area for research.

5.1.3   The nature of the 'NK receptor': candidate molecules and other molecules with poorly understood function

This aspect of NK cell biology remains highly enigmatic. Neither the structure of the molecule(s) mediating recognition of targets susceptible to NK cell killing nor its/their ligand(s) have been convincingly elucidated. It is known that NK cells neither rearrange TCR genes (90–93) nor express CD3 family member proteins on their cell surfaces (9, 93) although, as mentioned previously, the zeta chain is expressed (13). At present, one can only speculate whether the putative 'NK receptor' (NKR) should be a polymorphic or monomorphic structure. Certainly one would anticipate that less polymorphism would be required for an NKR than the TCR. The issue of the nature of the NKR is further compounded by the finding that upon treatment with IL-2, NK cells and even T-cells are able to lyse a much broader spectrum of targets. Presumably this could be due either to the expression of a new molecule or the expression of pre-existing molecules at a higher density. Since IL-2-activated NK and T-cells can mediate the virtually identical phenomenon of non-MHC-restricted activity, it is not clear if the molecules utilized by these two subsets of lymphocytes are similar, or if different cytolytic mechanisms are employed. An important question that might shed light on the nature of the NKR is how did NK cells and their receptors evolve? Despite the intense interest in the ability of NK cells to lyse tumour targets and the obvious therapeutic implications, it is unlikely that the tumouricidal activity of the cells was paramount in the evolutionary selection of NK cells and their molecules. Rather, the recognition of virally infected cells may have provided the selective advantage of this cell type. Studies utilizing human and murine NK cells, however, have identified not only their lytic capability as an important response to virally infected cells and microbial pathogens, but also their ability to secrete a wide range of cytokines. In addition, the ability of NK cells to recognize and regulate both leucocyte progenitors (30, 94) and more mature lymphocytes (95, 96) indicates that the broader functions of NK cells *in vivo* need to be reconsidered, when considering the structure of NKR. Although it is quite clear that CD3$^-$ LGLs, and for that matter IL-2-stimulated T-cells, may kill targets in a non-MHC-restricted manner, this does not exclude the possibility that expression of MHC molecules may influence the ability of these cells to kill targets. This may also have bearing on the nature of the putative NKR. This fascinating area has been recently reviewed (97).

Several approaches have been taken to delineate molecules of potential relevance to the recognition and lysis of tumour targets. The first, rather straightforward approach, has been to generate MABs capable of influencing NK cell-mediated killing. Using such an approach, a molecule involved in transmembrane signalling in NK cells has recently been cloned (98). Treatment of LGLs with this antibody results in killing of targets by reverse ADCC and also induces butoxycarbonyl-lysyl-tyrosine-esterase (BLT) release. This molecule is a 60 kDa disulphide-linked homodimer, which shares homology with asialoglycoprotein receptors, the low affinity Fcε receptor (FcεRII or CD23) and chondroitin sulphate proteoglycan core protein. In addition to the expression of this molecule by NK cells, the molecule appears to be expressed by neutrophils and to a lesser extent by T-cells. The ability of this molecule to confer cytolytic capability by transfection of its cDNA has not been reported, so the role of this molecule in this process remains unclear at present.

Imboden and co-workers have also generated two MABs against NK cells. One, raised against a 35 kDa protein on NK cells, has also shown the expression of this protein on other hematopoietic cells. A second molecule highlighted in this way is a 42 000 molecular weight surface molecular, which is expressed on IL-2 activated but not resting NK cells. This molecule is not expressed on either resting or IL-2-stimulated T-cells. Perturbation of both of these molecules induces increased phosphatidylinositol (PI) turnover and intracellular Ca^{2+} levels (99). Because of the lack of unique expression of these molecules on resting NK cells, it is unlikely that they are critically involved in NK-mediated cytolysis, although the latter molecule may be involved in LAK activity. GL138 is another MAB that selectively reacts with $CD2^+$, $CD3^-$, $CD16^+$, and $CD56^+$ LGLs (100, 101). However, the majority of $CD56^+$ cells do not react with this antibody. This MAB inhibits cytolysis of some tumour targets whilst the lysis of other targets is enhanced. Despite the potential importance of this molecule, it is unlikely to represent the NKR since it is not expressed on all NK cells.

In an attempt to define the receptors and structures involved in NK recognition, we have developed a MAB (designated MAB1) against NK target antigens on K562 cells that inhibits LGL binding and lysis (101). We subsequently developed an anti-idiotypic antibody (anti-ID) against MAB1 anticipating that it might recognize a NK surface molecule and aid in the identification of such a structure (102). The anti-ID antibody binds to NK cells and immunoprecipitates a complex of molecules (80 kDa and 150 kDa). In addition, it blocks conjugate formation and target cell lysis. Utilizing this anti-ID, we have cloned a novel gene from a NK-specific expressing library, which encodes a 150 kDa cell-surface peptide. This structure is a triggering molecule, since antisera blocks NK lysis, mediates reverse ADCC, induces NK cells to release cytokines (IFNγ, TNFα), induces BLT esterase release, and reacts specifically with freshly isolated and cloned NK cells. This molecule, however, is lacking on $CD3^+$ TCR α/β^+ or γ/δ^+ T-cell clones.

Recently, Evans and co-workers (103) reported that MABs directed against purified non-specific cytotoxic cells of the fish, *Ictalurus punctatus*, inhibited lysis of human NK target cells. All MABs were of the IgM isotype, and the inhibition of lysis occurred after preincubation of effector cells with the MAB. These MABs also significantly reduced NK activity against other human susceptible targets (MOLT-4, K562, Daudi) and murine target cells (P815, U937, YAC-1, and HL-60). The lack of a species-specific restriction in the effect of this antireceptor is surprising in the light of the species restriction of NK activity seen between man and mouse. Biochemical analysis of this effector moiety indicates that it is a heterodimer consisting of 38 kDa and 41 kDa proteins. This molecule has been proposed to be involved in the recognition and lysis of NK susceptible targets. The relationship of these 38–41 kDa proteins to other proposed NKRs is currently unclear and will require molecular cloning of this molecule.

5.2   Signal transduction

Upon activation of one of a variety of receptors on the NK cell surface, intracellular signals are generated that result in exocytosis of granule contents, lysis of targets, alteration in gene expression, and/or proliferation of NK cells. The biochemical steps leading to these events, however, are poorly understood. The data available come from two types of studies: pharmacological manipulations and measurement of second messengers initially elucidated in other receptor systems. Although these studies provide useful information, one limitation of this approach is that it tends to confirm existing paradigms. A receptor that may have particular relevance for receptor-mediated signalling in NK cells is the TCR on T-cells. Like so many receptors, the TCR appears to be coupled to PI hydrolysis, elevation of intracellular calcium concentrations, and activation of PKC activity in the plasma membrane (34, 35, 104–105). In addition, the TCR is coupled to a non-receptor protein tyrosine kinase (PTK) (34). How exactly these pathways interact is not yet entirely clear but TCR-dependent PI metabolism may be dependent upon both the function of a receptor-coupled PTK (106) and a protein tyrosine phosphatase activity (107). In addition, it appears that zeta has critical functions in coupling the TCR to these kinase pathways (54).

With respect to the exposure of NK cells to susceptible targets, this also results in a rise in inositol phosphate levels that is accompanied by an increase in intracellular-free calcium concentrations (108, 111). The rise in intracellular calcium levels is due to release of intracellular stores as well as influx of extracellular Ca^{2+}. As predicted, exposure of NK cells to tumour cells that resist NK cell-mediated lysis does not result in PI turnover or calcium flux. The interpretation of these studies has led to the notion that the putative NKR is coupled to PI hydrolysis and activation of a receptor-coupled phospholipase. Interestingly, although NK cells kill virally infected cells, these targets evidently do not induce PI turnover and calcium flux (112). This potentially fascinating finding suggests that distinct recognition molecules may be involved and/or that the biochemistry of NK cell activation initiated by virally infected targets is different from that mediated by tumour targets. Unfortunately, in the absence of any understanding of the molecular basis of recognition of these two targets, this remains merely an interesting phenomenon. Perturbation of NK cells by anti-FcγRIII MAB also induces PI turnover and calcium flux (108, 113, 114). Not unexpectedly, treatment of NK cells with a stimulatory combination of anti-CD2 MABs also results in a rise in intracellular calcium levels contributed to by intracellular, as well as extracellular sources (115). In many systems, direct pharmacological activation of PKC serves to uncouple receptors from PI turnover. Indeed, pretreatment of NK cells with PKC-activating phorbol esters inhibits the PI turnover and Ca^{2+} flux stimulated by occupation of the FcR. However, PI turnover and Ca^{2+} flux induced by NK-sensitive targets is not inhibited by pretreatment with phorbol esters, suggesting that the mechanisms which lead to the desensitization of these two receptors appear to be quite different (116).

Although the TCR has clearly been shown to couple to non-receptor PTK (34), little is known regarding the role of this class of kinases in NK receptor-mediated signalling. NK cells express a least two *src* family PTKs including pp56^{lck} (H. Young, unpublished observations) and pp58^{c-fgr} (117). Interestingly, in T-cells, pp56^{lck} has been shown to be physically as well as functionally associated with the accessory molecules, CD4 and CD8 (76). Since CD8 is expressed on a subset of NK cells, it is possible that CD8 may be also associated with pp56^{lck} in NK cells. However, if this is the case, the relevance of such an association is unclear. Since the ligand for CD8 is MHC Class I and NK cell-mediated lysis is not MHC-restricted, one might be led to hypothesize that the expression of CD8 is not crucial. This, however, would also not explain the function of pp56^{lck} in NK cells lacking CD8. Recently, it has been shown that pp55^{c-fcr} is associated with the granule fraction of neutrophils (118). Upon activation with chemotactic peptides, this PTK is translocated from the granule fraction to the plasma membrane together with other granule constituents. This raises the intriguing possibility that such a class of kinases may be important in regulating secretory events. We have recently demonstrated that the NK FcγR, like the TCR, appears to couple to a PTK (15). Perturbation of the FcR results in tyrosine phosphorylation of a number of substrates (Fig. 1.3) including the zeta chain, a molecule that is physically associated with this receptor. However, the identity of this PTK is not known at the present time.

It is not clear how activation of this pathway relates to the coupling of the FcR to PI turnover. These pathways may either be parallel or activation of a PTK may precede activation of phospholipase C (PLC). Recently, evidence has emerged to suggest that the latter may be correct. For some receptors, coupling to PLC may be relatively straightforward, in that a guanine nucleotide-binding protein is utilized. For other receptors, coupling appears to be much more complicated. For example, such receptors as the epidermal growth factor-receptor (EGFR) and the platelet-derived growth factor-receptor (PDGFR), are both coupled to PI turnover, but via mechanisms that involve isoforms of PLC being utilized by these receptors which become tyrosine phosphorylated following ligand binding reviewed in (120). In fact, it has been suggested that upon activation, the EGFR forms a signalling particle composed of PLCγ, the guanine nucleotide-activating protein (GAP, also a tyrosine phosphorylated substrate), phosphatidylinositol-3 kinase, and *raf* kinase. In addition, it now appears that cytoplasmic PTK of the *src* family also associates with the PDGFR following ligand binding (120). In the case of the TCR, a receptor

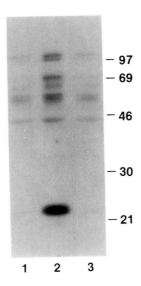

Fig. 1.3 Immunoblot analysis of tyrosine phosphorylated substrates in NK cells stimulated via FcγRIII (CD16). (Reprinted from O'Shea *et al.* (1991) (15).)

whose function seems to have considerable relevance to the FcR, activation of PLC appears to be dependent upon the function of PTK (121) as shown by the use of the PTK inhibitor, genestein. Moreover, activation of a PTK temporally precedes PI turnover (122). Additionally, however, guanine nucleotide-binding proteins including P21 c-*ras* may nonetheless have relevance in TCR-mediated signalling (123–126). As mentioned earlier, the mechanisms involved in TCR-mediated signalling are likely to have similarities to the FcR. Given the complexity of the systems, it is premature to do more than speculate regarding the coupling of these receptors to effector pathways. As will be discussed, the IL-2 receptor also appears to be coupled to a PTK, and so the role of PTK in NK-mediated signalling no doubt will be an area of vigorous future research.

Although protein kinases have received much attention as the biochemical means of signal transduction, this is not to say that these enzymes are the only means by which signalling occurs. Clearly, ion channels are important in the alteration of cellular function, although this mechanism has not been thoroughly evaluated in NK functions. It is clear, however, that Na^+ channel agonists, such as veratridice and batrachotoxin, inhibit cytotoxicity, and this effect is reversed by tetrodotoxin, a Na^+ channel antagonist (127). Despite this information, it is obvious that our understanding of signal transduction is in its infancy and remains rather phenomenological.

5.3   Distal effects of ligation of surface molecules

The distal effects that occur as a consequence of ligation of NK surface receptors include enhancement of lytic capacity, secretion of granule contents such as serine esterases, target lysis, alteration of gene expression, and proliferation. Recognition of targets or even membranes derived from target cells by NK cells induces a number of these events including IFNα, IFNγ, IL-1, and TNF secretion (10, 11). In addition, perturbation of the FcR has been shown to result in secretion of granule contents, lysis of targets, and the expression of IL-2Rα (CD25), cytokine genes (IFNγ, TNF), transferrin receptor (CD71), and 4F2 antigen (113, 128, 129). Unlike cytokine genes in the T-cells, induction of the IL-2 genes requires *de novo* protein synthesis for transcription. In contrast, genes induced in NK cells by ligation of the FcR evidently do not require *de novo* protein synthesis. This suggests that gene regulation in the two related cell types may have marked differences. Interestingly, perturbation of FcγRIII does not induce NK cells to proliferate.

Engagement of the CD2 molecule on T-cells is a mitogenic signal and has been suggested to be an alternate pathway of antigen-independent activation. In NK cell clones (130) perturbation of CD2 by combinations of anti-CD2 MABs increases cytotoxicity. Combinations of anti-CD2 MAB result in increased formation of conjugates, enhancement or inhibition of lytic activity, LAK activity, and exocytosis of cytolytic granules (130). These results indicate that the anti-CD2 interactions can modulate the cytotoxic potential of the effector cell and trigger cytolytic machinery resulting in target cell destruction. Anti-CD2 antibodies also induce transcription and translation of the IFNγ gene (131). These data provide evidence that the CD2 molecule may be an important physiological regulator of functions of CD3$^-$LGLs and also provide impetus for further investigation of the role of this molecule in both the cytotoxic and noncytotoxic functions of these lymphocytes.

As mentioned above, the cell biology of NK granules and the secretory events involved in the release of such lytic molecules as the pore-forming protein and TNFα have become increasingly well understood. It is beyond the scope of the present review to do justice to this exciting and rapidly expanding field.

5.3.1   *Pharmacological studies*

Although studies that measure the generation of second messengers are useful in delineating the pathways employed by various receptors in transmembrane signalling, they do not necessarily address the issue of the function of the various pathways or what steps are necessary or sufficient for the induction of various events. In other words, second messenger studies do not necessarily provide information regarding the consequence of activation of different pathways. Pharmacological studies, despite their limitations, can begin to address the issue of the functional relevance of these pathways. The studies enumerated in the preceding section indicate that occupancy of the FcR, CD2, and putative NKR result in PI turnover and calcium flux. One would anticipate then that this should result in the activation of PKC. Thus, the pharmacological activation of PKC should induce some of the changes associated with ligand occupancy of these receptors. There is evidence from a number of studies that direct activation of PKC with phorbol esters can induce many of these effects. For example, treatment of LGLs with phorbol myristate acetate increases NK cell-mediated cytolytic activity against K562 and induces LAK activity (132). In addition, such phorbol esters induce IFNγ production (133). This latter finding differs considerably from the response of T-cells to phorbol esters; where treatment with such an agent alone is generally not sufficient to induce the expression of the IL-2 or IFNγ genes. This supports the hypothesis that some mechanism involved in signal transduction or in the regulation of gene expression is quite different between these two lymphocyte populations (133). However, this is not a completely unexpected finding since T-cells and NK cells also differ in their response to IL-2. Although IL-2 alone induces IFNγ secretion, IFNγ expression in T-cells requires additional second signals. Diacyglycerols and phorbol esters synergize with Ca^{2+} ionophore (134) to induce proliferation of LGLs, CD25 (IL-2Rα) and transferrin receptor (CD71) expression. These findings support an important role for PKC in the pathways leading to the activation of NK cells but suggest that intracellular signalling may be somewhat different in T-cells and LGLs.

An alternative way of establishing a role for PKC in these various events is to utilize PKC inhibitors. This approach, although potentially useful, is fraught with more difficulties than the former approach, because most inhibitors of PKC are not specific for this enzyme alone. Whereas the IC50 of a potent drug, such as staurosporine for PKC, is 60 nM, the IC50 of this drug for other kinases, such as the EGFR kinase and S6 kinase, are not much greater (135). Thus, these agents might inhibit kinases other than PKC that are also critical in the function of NK cells. For example, inhibition of myosin light chain kinase inhibits NK activity (136). With these caveats in mind, it has been shown that H-7, retinol, and staurosporine are potent inhibitors of basal NK cell cytolytic activity, ADCC, IL-2-induced IFNγ production, and LAK activity. This inhibition can be reversed by removal of pharmacological agents and reculturing with IL-2 (137–140).

Unfortunately, no pharmacological PTK activator exists at the present time for use in an approach analogous to the activation of PKC with phorbol esters. However, a number of PTK inhibitors are now available including genestein and tyrphostins (141) and experiments are underway to determine the effect of these agents on NK cell function.

5.3.2   Molecules involved in the enhancement of NK cell function

The *in vivo* activation of NK activity by bacterial products was one of the first observations to link cytokines with the immunological modulation of NK cells. These initial observations led to further studies involving the modulation of NK cell lytic function by IFNα/β. Table 1.5 summarizes the

Table 1.5   Effects of recombinant cytokines on NK activity

Cytokine	*In vitro*	*In vivo*
IFNα/β	↑	↑
IFNγ	±	↑
IL-1	±	↑
IL-2	↑	↑
IL-3	→	↓
IL-4	↓	?
IL-5	→	?
IL-6	↑	?
IL-7	↑	?

↑, increases activity; →, no change in activity; ↓, decreases activity; ±, marginal increase in activity.

effects of various interferons as well as interleukins 1–7 on NK cells. It was found that exposure to IFNs causes a dose-dependent activation of NK activity and that this augmentation has similar characteristics to that seen following viral infection. These findings led to the first production and use of a recombinant cytokine, IFNα. IFNα/β has been shown to be the active agent in many *in vitro* and *in vivo* preparations which modulated NK cell function. Recombinant IFNα/β has effects at all levels of the lytic process enhancing, binding, triggering, and lysis. Phase I clinical trials with IFNα or IFNβ have demonstrated that NK activity peaks at 12 to 24 hours following IFN injection (142). Although IFNγ has very modest ability to modulate NK activity *in vitro*, it substantially increases the cytolytic and secretory activity of NK cells in both murine (11, 143) and human systems (144, 145).

Of the recombinant cytokines surveyed, however, IL-2 is to date the most potent activator of NK activity (see below). IL-3 (146, 147) and IL-5 (J. J. O'Shea and J. R. Ortaldo, unpublished observations) are inactive, while IL-4 can be inhibitory and IL-1 appears to modestly modulate NK activity but does so indirectly (148, 149). The recently described interleukins, IL-6 and IL-7, modestly augment NK activity *in vitro* (150, 151). However, most of the ILs have not been adequately tested in animals or in humans to verify their *in vivo* regulatory ability.

Activation of CD3⁻ LGLs by IL-2 results in the enhancement of lytic activity within minutes, although maximal cytolytic capacity is observed by 24 hours (23, 146). As discussed earlier, IL-2 treatment also broadens the targets that are susceptible to NK-mediated killing, that is it induces LAK activity in NK cells (152, 153). In addition to modulation of cytotoxic activities, IL-2 is capable of altering the expression of a number of receptors and cytokine genes that are transcribed and expressed on the surface of CD3⁻LGLs (IL-2Rα and HLA-Dr), or secreted by the cells (IFNγ) (133, 154, 155).

Early studies revealed that CD3⁻lymphocytes require higher doses of IL-2 than T-cells to induce maximal proliferation. Whereas activated T-cells or T-cell lines require 1–25 units/ml of this cytokine to induce maximal proliferation, CD3⁻LGLs require 100–500 units/ml for a similar effect (154, 156). These data are consistent with other reports (157, 158) indicating that more high affinity receptors were expressed on activated T-cells than on CD3⁻LGLs. However, in contrast to T-cells, which require more than one stimulus for optimal activation [reviewed in (79)], CD3⁻LGLs will respond to IL-2 alone. This finding suggested that, in the basal state, LGLs might express an IL-2-binding protein that is distinct from the IL-2R (the TAC molecule), which had only recently been described at the time. Recently, this problem has been solved by the discovery in several laboratories that a 75 000 mol. wt. (IL-2Rβ) IL-2-binding protein is expressed on LGLs (157–161). Further studies have demonstrated that CD3⁺ T-cells also express p75 IL-2Rβ, but that the receptor level per cell is considerably less than on NK cells (162) and, indeed, is only detectable by functional analysis (163). Following stimulation, the IL-2Rα gene is induced and confers high affinity ligand binding.

NK cells possess an IL-2Rβ receptor which can generate signals to induce functional changes observed within minutes, such as activation of NK cell lytic activity. Alternatively, activation can occur which requires transcriptional activity, gene transcription, and translation, such as that of the IFNγ gene. The signals and pathways involved in these biochemical pathways are of considerable interest. Initial efforts aimed at dissecting out the signals involved in IL-2Rβ-mediated events have involved pharmacological studies and assessment of phosphorylation events. Although pharmacological agents can be broad and often non-specific, they do allow the rapid analysis of events. One major pathway utilized by IL-2Rβ has been defined by studying the effects of both pharmacological activation of PKC, and PKC inhibitors. Agents that activate PKC, such as phorbol esters and phorbol derivatives, augment NK activities and induced NK cells to exhibit an activated state, such as their enhanced secretion of IFNγ. A variety of

PKC inhibitors (e.g. H7, retinal, and staureosporin) block IL-2-dependent NK cell activation (137, 138). Taken together with the stimulatory effects of phorbol esters, these data suggest that IL-2Rβ signalling events may be dependent upon PKC. In addition, however, tyrosine phosphorylation of specific substrates is also induced in both T- and NK cells upon activation by IL-2 indicating that the IL-2R is coupled to a PTK (163–166). In studies where tyrosine phosphorylated proteins were examined in both CD3$^+$ and CD3$^-$LGLs, the IL-2Rβ but not the IL-2Rα subunit was found to be sufficient to induce these events (164). In CD3$^-$LGLs only expressing IL-2Rβ, rapid phosporylation of both serine and tyrosine residues on a variety of protein substrates was detected. Interestingly, the IL-2Rβ chain itself is a tyrosine phosphorylated substrate (167, 168). Again, it is not entirely clear how the activation of a PTK and PKC relate to another. It is possible that, as for the TCR, IL-2R-mediated activation of PKC may be dependent upon a PTK. In reality, it is quite likely that the process of signalling from the plasma membrane to the nucleus entails a cascade of kinases, some of which are PTK and some of which are serine threonine kinases. There is data to support such a hypothesis. Upon stimulation of T-cells, tyrosine phosphorylation of at least two serine threonine kinases occurs: *raf* kinase (169), and *cdc* 2 kinase (G. A. Evans and W. L. Farrar, personal communication). The latter kinase has received enormous attention recently and is critically involved in the process of mitosis (170) and may phosphorylate among other substrates the retinoblastoma (Rb) protein (171). Interestingly, it has recently been shown in T-cells that Rb is phosphorylated in response to stimulation with IL-2 (G. A. Evans and W. L. Farrar, personal communication). The relevance of these findings to NK cell function is not yet known.

6   Summary

Considerable progress has been made in recent years in moving the investigation of NK cells from the phenomenological study of an activity in peripheral blood cells to delineating the functions of molecules central to the function of these cells. Remarkable advances have been made in understanding the structure and function of two key receptors of CD3$^-$ LGLs, FcγRIII, and the IL-2R. The biochemical means by which these receptors exert their effect on secretory processes and gene transcription are less well understood but the enzymes in the pathway, principally kinases, are now being characterized. The 'Holy Grail' of NK cell biology, the 'NK receptor', remains elusive but a number of NK-specific membrane proteins are now being cloned and it is hoped that at least one of these molecules will shed light on the precise mechanism by which NK cells recognize tumour- and virally infected targets. Discoveries such as these will have considerable relevance as an immunological principle of recognition of self and non- or modified self. In addition, a better understanding of NK and LAK cell activities is likely to have a definite therapeutic impact. It is for both these reasons that the NK cells/CD3⁻LGLs warrant the attention they now receive.

References

1 Rosenberg, E. B., Herberman, R. B., Levine, P. H., Halterman, R. H., McCoy, J. L., and Wunderlich, J. R. (1972). Lymphocyte cytoxicity reactions to leukemia-associated antigens in identical twins. *Int. J. Cancer*, **9**, 648–54.

2 Oldham, R. K., Siewarski, D., McCoy, J. L., Plata, E. J., and Herberman, R. B. (1973). Evaluation of cell-mediated cytotoxicity assay utilizing ^{125}iododeoxyuridine labelled tissue culture target cells. *Natl. Cancer Inst. Monogr.*, **37**, 49–58.

3 McCoy, J. L., Herberman, R. B., Rosenberg, E. C., Donnelly, F. C., Levine, P. H., and Alford, C. (1973). ^{51}Chromium release assay for cell-mediated cytotoxicity of human leukemia and lymphoid tissue-culture cells. *Natl. Cancer Inst. Monogr.*, **37**, 59–68.

4 Takasugi, M., Mickey, M. R., and Terasaki, P. I. (1973). Quantitation of the microassay for cell-mediated immunity through electronic image analysis. *Natl. Cancer Inst. Monogr.*, **37**, 77–84.

5 Kiessling, R., Klein, E., and Wigzell, H. (1975a). "Natural" killer cells in the mouse. I. Cytotoxic cells with specificity for mouse Moloney leukemia cells. Specificity and distribution according to genotype. *Eur. J. Immunol.*, **5**, 112–21.

6 Timonen, T., Ortaldo, J. R., and Herberman, R. B. (1981) Characteristics of human large granular lymphocytes and relationship to natural killer and K cells. *J. Exp. Med.*, **153**, 569–82.

7 Hercend, T., Reinherz, E. L., Meuer, S., Schlossman, S. F., and Ritz, J. (1983). Phenotypic and functional heterogeneity of human cloned natural killer cell lines. *Nature*, **301**, 158–60.

8 Lanier, L. L., Le, A. M., Civin, C. I., Liken, M. R., and Phillips, J. G. (1986). The relationship of CD16 (Leu-11) and Leu-19 (NKH-1) antigen expression on human peripheral blood NK cells and cytotoxic T lymphocytes. *J. Immunol.*, **136**, 4480–6.

9 Ortaldo, J. R., Sharrow, S. O., Timonen, T., and Herberman, R. B. (1981). Determination of surface antigens on highly purified human NK cells by flow cytometry with monoclonal antibodies. *J. Immunol.*, **127**, 2401–9.

10 Ortaldo, J. R. and Herberman, R. B. (1984). Heterogeneity of natural killer cells. In *Annual Reviews of Immunology* (ed. W. E. Paul, C. G.

Fathman, and H. Metzger), pp. 359–94. Annual Reviews Inc., Palo Alto, CA.

11 Trinchieri, G. (1989). Biology of natural killer cells. In *Advances in Immunology* (ed. F. J. Dixon), pp. 187–303. Academic Press, San Diego, CA.

12 Welsh, R. M. (1985). Natural killer cells and interferon. *CRC Crit. Rev. Immunol.*, **5**, 55–93.

13 Anderson, P., Caligiuri, M., Ritz, J., and Schlossman, S. F. (1989). CD3-negative natural killer cells express zeta TCR as part of a novel molecular complex. *Nature*, **341**, 159–62.

14 Lanier, L. L., Yu, G., and Phillips, J. H. (1989). Co-association of CD3 zeta with a receptor (CD16) for IgG Fc on human natural killer cells. *Nature*, **342**, 803–5.

15 O'Shea, J. J., Weissman, A. M., Kennedy, I. C. S., and Ortaldo, J. R. (1991). Engagement of the NK cell IgG Fc receptor results in tyrosine phosphorylation of the ζ chain. *Proc. Natl. Acad. Sci. U.S.A.*, **88**, 350–4.

16 Fitzgerald-Bocarsly, P., *et al.* (1989). A definition of natural killer cells. In *Natural killer cells and host defense* (ed. E. W. Ades and C. Lopez), p. xiii. Karger, New York.

17 Luini, W., Boraschi, S., Alberti, A., Alloetti, A., and Tagliabue, A. (1981). Morphological characterization of a cell population responsible for natural killer activity. *Immunology*, **43**, 663–9.

18 Kumagai, K., Itoh, K., Suzuki, R., Hinuma, S., and Saitoh, F. (1982). Studies of murine large granular lymphocytes. I. Identification a effector cells in NK and K cytotoxicities. *J. Immunol.*, **129**, 388–94.

19 Reynolds, C. W., Timonen, T., and Herberman, R. B. (1981). Natural killer (NK) cell activity in the rat. I. Isolation and characterization of the effector cells. *J. Immunol.*, **127**, 6351–62.

20 Reynolds, C. W., Sharrow, S. O., Ortaldo, J. R., and Herberman, R. B. (1981). Natural killer activity in the rat. II. Analysis of surface antigens on LGL by flow cytometry. *J. Immunol.*, **127**, 2204–8.

21 Nagler, A., Lanier, L. L., Cwirla, S., and Phillips, J. H. (1989). Comparative studies of human FcRIII- positive and negative natural killer cells. *J. Immunol.*, **143**, 3183–91.

22 Itoh, K., Tilden, A. B., Kumagai, K., and Balch, C. M. (1985) Leu-11$^+$ lymphocytes with natural killer (NK) activity are precursors of recombinant interleukin 2 (rIL-2)-induced activated killer (AK) cells. *J. Immunol.*, **134**, 802–7.

23 Ortaldo, J. R., Mason, A., and Overton, R. (1986). Lymphokine-activated killer (LAK) cells: Analysis of progenitors, and effectors. *J. Exp. Med.*, **164**, 1193–1205.

24 Phillips, J. H. and Lanier, L. L. (1986). Dissection of the lymphokine-activated killer phenomenon: relative contribution of peripheral blood natural killer phenomenon: relative contribution of peripheral blood

natural killer cells and T lymphocytes to cytolysis. *J. Exp. Med.*, **136**, 1579–85.

25 Phillips, J. H., Takeshita, T., Sugamura, K., and Lanier, L. L. (1989). Activation of natural killer cells via the p75 interleukin 2 receptor. *J. Exp. Med.*, **170**, 291–6.

26 Ortaldo, J. R., Frey, J., Takeshita, T., and Sugamura, K. (1990). Regulation of CD3⁻ lymphocyte function with an antibody against the IL-2β chain receptor: modulation of NK and LAK activity and production of IFNγ. *Eur. Cytokine Net.*, **1**, 27– 34.

27 Murphy, J. W. and McDaniel, D. O. 1982). *In vitro* reactivity of natural killer (NK) cells against *Cryptococcus neoformans*. *J. Immunol.*, **128**, 1577–83.

28 Pohajdak, B., Gomez, J. L., Wilkins, J. A., and Greenberg, A. H. (1984). Tumor-activated NK cells trigger monocyte oxidative metabolism. *J. Immunol.*, **133**, 2430–6.

29 Fitzgerald, P. A. and Lopez, C. (1986). Natural killer cells active against viral, bacterial, protozoan, and fungal infections. In *Immunobiology of natural killer cells* (ed. E. Lotzova and R. B. Herberman), pp. 107–132. CRC Press, Boca Raton, FL.

30 Cudowicz, G. and Hochman, P. S. (1979). Do natural killer cells engage in regulated reactions against self to ensure homeostatsis? *Immunol. Rev.*, **44**, 13–41.

31 Lotzova, E. (1986). NK cell role in regulation of the growth and functions of hemopoietic and lymphoid cells. In *Immunobiology of natural killer cells* (ed. E. Lotzova and R. B. Herberman), pp. 89–106. CRC Press, Boca Raton, Fl.

32 Trincheri, G. and Perussia, B. (1984). Biology of disease. Human natural killer cells: biologic and pathologic aspects. *Lab Invest.*, **50**, 489–513.

33 Scala, G., Djeu, J. Y., Allavena, P., Kasahara, T., Ortaldo, J. R., Herberman, R. B., and Oppenheim, J. J. (1986). Cytokine secretion and noncytotoxic functions of human large granular lymphocytes. In *Immunobiology of natural killer cells, vol. II* (ed. E. Lotzova and R. B. Herberman), pp. 133–44. CRC Press, Boca Raton, Fl.

34 Weissman, A. M., Bonifacino, J. S., Klausner, R. D., Samelson, L. E. S., and O'Shea, J. J. (1989). T cell antigen receptor structure, assembly, and function. *Year Immunol.*, **4**, 74–93.

35 Goldsmith, M. A. and Weiss, A. (1988). New clues about T-cell antigen receptor complex function. *Immunol. Today*, **9**, 220–2.

36 Acuto, O., *et al.* (1985). The human T cell receptor. *J. Clin. Immunol.*, **5**, 141–50.

37 Fanger, M. W., Shen, L., Graziano, R. F., and Guyre, P. M. (1989). Cytotoxicity mediated by human Fc receptors for IgG. *Immunol. Today*, **10**, 92–9.

38 Kinet, J. P. (1989). Antibody–cell interactions: Fc receptors. *Cell*, **57**, 351–4.

39 Unkeless, J. C. (1989). Human Fc Gamma receptors. *Curr. Opin. Immunol.*, **2**, 63–7.

40 Simmons, D. and Seed, B. (1988). The Fc gamma receptor of natural killer cells is a phospholipid-linked membrane protein. *Nature*, **333**, 568–70.

41 Lanier, L. L., Ruitenberg, J. J., and Phillips, J. H. (1988). Functional and biochemical analysis of CD16 antigen on natural killer cells and granulocytes. *J. Immunol.*, **141**, 3478–85.

42 Lanier, L. L., Phillips, J. H., and Testi, R. (1989). Membrane anchoring and spontaneous release of CD16 (FcR III) by natural killer cells and granulocytes. *Eur. J. Immunol.*, **19**, 775–8.

43 Ueda, E., Kinoshita, T., Nojima, J., Inoue, K., and Kitani, T. (1989). Different membrane anchors of Fc gamma RIII (CD16) on K/NK-lymphocytes and neutrophils. Protein- vs lipid-anchor. *J. Immunol.*, **143**, 1274–7.

44 Edberg, J. C., Redecha, P. B., Salmon, J. E., and Kimberly, R. P. (1989). Human Fc gamma RIII (CD16). Isoforms with distinct allelic expression, extracellular domains, and membrane linkages on polymorphonuclear and natural killer cells. *J. Immunol.*, **143**, 1642–9.

45 Selvaraj, P., Carp'en, O., Hibbs, M. L., and Springer, T. A. (1989). Natural killer cell and granulocyte Fc gamma receptor III (CD16) differ in membrane anchor and signal transduction. *J. Immunol.*, **143**, 3283–8.

46 Ravetch, J. V. and Perussia, B. (1989). Alternative membrane forms of Fc gamma RIII (CD16) on human natural killer cells and neutrophils. Cell type-specific expression of two genes that differ in single nucleotide substitutions. *J. Exp. Med.*, **170**, 481–97.

47 Lanier, L. L., Cwirla, S., Yu, G., Testi, R., and Phillips, J. H. (1989). Membrane anchoring of a human IgG Fc receptor (CD16) determined by a single amino acid. *Science*, **246**, 1611–13.

48 Kurosaki, T. and Ravetch, J. V. (1989). A single amino acid in the glycosyl phosphatidylinositol attachment domain determines the membrane topology of Fc gamma RIII. *Nature* **342**, 805–7.

49 Zeger, D. L., Hogarth, P. M., and Sears, D. W. (1990). Characterization and expression of an Fc gamma receptor cDNA cloned from rat natural killer cells. *Proc. Nat. Acad. Sci. U.S.A.*, **87**, 3425–9.

50 Miller, L., Blank, U., Metzger, H., and Kinet, J. P. (1989). Expression of high-affinity binding of human immunoglobulin E by transfected cells. *Science*, **244**, 334–7.

51 Anderson, P., Caligiuri, M., O'Brien, C., Manley, T., Ritz, J., and Schlossman, S. F. (1990). Fc gamma receptor type III (CD16) is included in the zeta NK receptor complex expressed by human natural killer cells. *Proc. Natl. Acad. Sci. U.S.A.*, **87**, 2274–8.

52 Ra, C., Jouvin, M. H., Blank, U., and Kinet, J. P. (1989). A macrophage Fc gamma receptor and the mast cell receptor for IgE share an identical subunit. *Nature*, **341**, 752–4.

53 Hibbs, M. L., *et al.* (1989). Mechanisms for regulating expression of membrane isoforms of Fc gamma RIII (CD16). *Science*, **246**, 1608–11.

54 Ashwell, J. D. and Klausner, R. D. (1990). Genetic and mutational analysis of the T-cell antigen receptor. *Ann. Rev. Immunol.*, **8**, 139–67.

55 Orloff, D. G., Ra, C. S., Frank, S. J., Klausner, R. D., and Kinet, J-P. (1990). Family of disulfide-linked dimers containing the zeta and eta chains of the T-cell receptor and the gamma chain of Fc receptors. *Nature*, **347**, 189–91.

56 Perussia, B., *et al.*, (1989). Murine natural killer cells express functional Fc gamma receptor II encoded by the Fc gamma R alpha gene. *J. Exp. Med.*, **170**, 73–86.

57 Moingeon, P., Stebbins, C. C., D'Adamio, L., Lucich, J., and Reinherz, E. L. (1990). Human natural killer cells and mature T lymphocytes express identical CD3ζ subunits as defined by cDNA cloning and sequence analysis. *Eur. J. Immunol.*, **20**, 1741–5.

58 Yagita, H., Okumura, K., and Nakauchi, H. (1988). Molecular cloning of the murine homologue of CD2. Homology of the molecule to its human counterpart T11. *J. Immunol.*, **140**, 1321–6.

59 Scott, C. F., *et al.* (1989). Activation of human cytolytic cells through CD2/T11. Comparison of the requirements for the induction and direction of lysis of tumor targets by T cells and NK cells. *J. Immunol.*, **142**, 4105–12.

60 Webb, D. S. A., Shimizu, Y., VanSeventer, G. A., Shaw, S., and Gerrard, T. L. (1990). LFA-3, CD44, and CD45: Physiologic triggers of human monocyte TNF and IL-1 release. *Science*, **249**, 1295–8.

61 Shevach, E. M. (1989). Accessory molecules. In *Fundamental immunology* (ed. W. Paul), pp. 413–44. Raven, New York.

62 Hart, M. K., Kornblutch, J., Main, E. K., Spear, B. T., Taylor, J., and Wilson, D. B. (1987). Lymphocyte function-associated antigen 1 (LFA-1) and natural killer (NK) cell activity: LFA-1 is not necessary for all killer: target cell interactions. *Cell. Immunol.*, **109**, 306–17.

63 Timonen, T., Patarroyo, M., and Gahmberg, C. G. (1988). CD11a-c/CC18 and GP84 (LB-2) adhesion molecules on human large granular lymphocytes and their participation in natural killing. *J. Immunol.*, **141**, 1041– 6.

64 van Kooyk, Y., van de Wiel-van Kemenade, P., Weder, P., Kuijpers, T. W., and Figdor, C. G. (1989). Enhancement of LFA-1-mediated cell adhesion by triggering through CD2 or CD3 on T lymphocytes. *Nature*, **342**, 811–13.

65 Dustin, M. L. and Springer, T. A. (1989). T-cell receptor cross-linking transiently stimulates adhesiveness through LFA-1. *Nature*, **341**, 619–24.

66 Springer, T. A. (1990). Adhesion receptors of the immune system. *Nature*, **346**, 425–34.

67 Ramos, O. F., Kai, C., Yefenof, E., and Klein, E. (1988). The elevated natural killer sensitivity of targets carrying surface-attached C3 fragments require the availability of the iC3b receptor (CR3) on the effectors. *J. Immunol.*, **140**, 1239–43.

68 Ramos, O. F., Patarroyo, M., Yefenof, E., and Klein, E. (1989). Requirement of leukocyte cell adhesion molecules (CD111a-c/CD18) in the enhanced NK lysis of iC3b-opsonized targets. *J. Immunol.*, **142**, 4100–4.

69 Yefenof, E., Benizri, R., Reiter, Y., Klein, E., and Fishelson, Z. (1990). Potentiation of NK cytotoxicity by antibody-C3b/iC3b heteroconjugates. *J. Immunol.*, **144**, 1538–43.

70 Berger, M., O'Shea, J. J., Cross, A. S., Chused, T. M., Brown, E. J., and Frank, M. M. (1984). Isolated human neutrophil leukocytes respond to activating stimuli change Fc receptor expression. *J. Clin. Invest.*, **74**, 1566–71.

71 Lanier, L. L., Testi, R., Bindl, J., and Phillips, J. H. (1989). Identity of Leu-19 (CD56) leukocyte differentiation antigen and neural cell adhesion molecule. *J. Exp. Med.*, **169**, 2233–8.

72 Schwarz, R. E. and Hiserodt, J. C. (1988). The expression and functional involvement of laminin-like molecules in non-MHC restricted cytotoxicity by human Leu-19$^+$/CD3$^-$ natural killer lymphocytes. *J. Immunol.*, **141**, 3318–23.

73 Santoni, A. *et al.* (1989). Rat natural killer cells synthesize fibronectin. Possible involvement in the cytotoxic function. *J. Immunol.*, **143**, 2415–21.

74 Perussia, B., Fanning, V., and Trinchieri, G. (1983). A human NK and K cell subset shares with cytotoxic T cells expression of the antigen recognized by antibody OKT8. *J. Immunol.*, **131**, 223–31.

75 Bolhuis, R. L., Roozemond, R. C., and van de Griend, R. J. 1986). Induction and blocking of cytolysis in CD2+, CD3−NK and CD2+, CD3+ cytotoxic T lymphocytes via CD2 50 KD sheep erythrocyte receptor. *J. Immunol.*, **136**, 3939–44.

76 Rudd, C. E., Anderson, P. M., Morimoto, C., Streuli, M., and Schlossman, S. F. (1989). Molecular interactions, T-cell subsets and a role of the CD4/CD8: p56^{lck} complex in human T-cell activation. *Immunol. Rev.*, **111**, 225–66.

77 Charbonneau, H., Tonks, N. K., Walsh, K. A., and Fischer, E. H. (1988). The leukocyte common antigen (CD45): A putative receptor-linked protein tyrosine phosphatase. *Proc. Natl. Acad. Sci. U.S.A.*, **85**, 7182–6.

78 Charbonneau, H., *et al.* (1989). Human placenta protein-tyrosine-phosphatase: Amino acid sequence and relationship to a family of a receptor-like proteins. *Proc. Natl. Acad. Sci. U.S.A.*, **86**, 5252–6.

79 Tonks, N. K., Charbonneau, H., Diltz, C. D., Fischer, E. H., and Walsh, K. A. (1988). Demonstration that the leukocyte common antigen CD45 is a protein tyrosine phosphatase. *Biochemistry*, **27**, 8695–701.

80 Tonks, N. K., Diltz, C. D., and Fischer, E. H. (1990). CD45, an integral membrane protein tyrosine phosphatase. Characterization of enzyme activity. *J. Biol. Chem.*, **265**, 10674–80.

81 Saga, Y., Lee, J. S., Saraiya, C., and Boyse, E. A. (1990). Regulation of alternative splicing in the generation of isoforms of the mouse LY-5 (CD45) glycoprotein. *Proc. Natl. Acad. Sci. U.S.A.*, **87**, 3728–32.

82 Ledbetter, J. A., Tonks, N. K., Fischer, E. H., and Clark, E. A. (1988). CD45 regulates dignal transduction and lymphocyte activation by specific association with receptor molecules on T or B cells. *Proc. Natl. Acad. Sci. U.S.A.*, **85**, 8628–32.

83 Kiener, P. A. and Mittler, R. S. (1989). CD45-protein tyrosine phosphatase cross-linking inhibits T cell receptor CD3-mediated activation in human T cells. *J. Immunol.*, **143**, 23–8.

84 Deusch, K., *et al.* (1990). Differential regulation of Ca^{2+} mobilization in human thymocytes by coaggregation of surface molecules. *J. Immunol.*, **144**, 2851–8.

85 Volarevic, S., Burns, C. M., Sussman, J. J., and Ashwell, J. D. (1990). Intimate association of Thy-1 and the T-cell antigen receptor with the CD45 tyrosine phosphatase. *Proc. Natl. Acad. Sci. U.S.A.*, **87**, 7085–9.

86 Mustelin, T., Coggeshall, K. M., and Altman, A. (1989). Rapid activation of the T-cell tyrosine protein kinase pp56lck by the CD45 phosphotyrosine phosphatase. *Proc. Natl. Acad. Sci. U.S.A.*, **86**, 6302–3.

87 Ostergaard, H. L. *et al.* (1989). Expression of CD45 alters phosphorylation of the lck-encoded tyrosine protein kinase in murine lymphoma T-cell lines. *Proc. Natl. Acad. Sci. U.S.A.*, **86**, 8959–63.

88 Ostergaard, H. L. and Trowbridge, I. S. (1990). Coclustering CD45 with CD4 or CD8 alters the phosphorylation and kinase activity of p56lck. *J. Exp. Med.*, **172**, 347–50.

89 Mustelin, T. and Altman, A. (1990). Dephosphorylation and activation of the T cell tyrosine kinase pp56lck by the leukocyte common antigen (CD45). *Oncogene*, **5**, 809–13.

90 Reynolds, C. W., Bonyhadi, M., Herberman, R. B., Young, H. A., and Hedrick, S. M. (1985). Lack of gene rearrangement and mRNA expression of the beta chain of the T cell receptor in spontaneous rat large granular lymphocyte leukemia lines. *J. Exp. Med.*, **161**, 1249–54.

91 Lanier, L. L., Cwirla, S., Federspiel, N., and Phillips, J. H. (1986). Human natural killer cells isolated from peripheral blood do not rearrange T cell antigen receptor beta chain genes. *J. Exp. Med.*, **163**, 209–14.

92 Biron, C. A., van den Elsen, P., Tutt, M. M., Medveczky, P., Kumar,

V., and Terhorst, C. (1987). Murine natural killer cells stimulated *in vivo* do not express the T cell receptor alpha, beta, gamma, T3 delta, or T3 epsilon genes. *J. Immunol.*, **139**, 1704–10.

93 Ritz, J., *et al.* (1985). Analysis of T cell receptor gene rearrangement and expression in human natural killer cell clones. *Science*, **228**, 1540–7.

94 Kumar, V., Luevano, E., and Bennet, M. (1979). Hybrid resistance to EL-4 lymphoma cells. I. Characterization of natural killer cells that lyse EL-4 cells and their distinction from marrow-dependent natural killer cells. *J. Exp. Med.*, **150**, 531–9.

95 Michael, A., Hackett, J. J., Bennett, M., Kumar, V., and Yuan, D. (1989). Regulation of B lymphocytes by natural killer cells. Role of IFN-gamma. *J. Immunol.*, **142**, 1095–101.

96 Robles, C. P. and Pollack, S. B. (1989). Asialo-GM1+ natural killer cells directly suppress antibody-producing B cells. *Nat. Immun. Cell Growth Regul.*, **8**, 209–22.

97 Hogland, P., *et al.* (1988). Natural resistance against lymphoma grafts conveyed by H-2Dd transgene to C57BL mice. *J. Exp. Med.*, **186**, 1469–74.

98 Giorda, R., Rudert, W. A., Vavassori, C., Chambers, W. H., Hiserodt, J. C., and Trucco, M. (1990). NKR-P1, a signal transduction molecule on natural killer cells. *Science*, **249**, 1298–1300.

99 Imboden, J. B., Eriksson, E. C., McCutcheon, M., Reynolds, C. W., and Seaman, W. E. (1989). Identification and characterization of a cell-surface molecule that is selectively induced on rat lymphokine-activated killer cells. *J. Immunol.*, **143**, 3100–3.

100 Moretta, A., *et al.* (1989). Surface molecules involved in the activation and regulation of T or natural killer lymphocytes in humans. *Immunol. Rev.*, **111**, 145–75.

101 Moretta, A., *et al.* (1990). A novel surface antigen expressed by a subset of human CD3⁻ CD16⁺ natural killer cells. Role in cell activation and regulation of cytolytic function. *J. Exp. Med.*, **171**, 695–714.

102 Ortaldo, J. R., Kantor, R., Segal, D., Bolhuis, R. H., and Bino, T. (1989). Identification of a proposed NK receptor. In *Natural killer cells* (ed. E. W. Ades and C. Lopez), pp. 221–6. Karger, Basel.

103 Evans, D. L., Jaso-Friedmann, L., Smith, E. E., Jr., St. John, A., Koren, H. S., and Harris, D. T. (1988). Identification of a putative antigen receptor on fish nonspecific cytotoxic cells with monoclonal antibodies. *J. Immunol.*, **141**, 324–32.

104 Berridge, M. J. and Irvine, R. F. (1989). Inositol phosphates and cell signalling. *Nature*, **21**, 197–205.

105 Kikkawa, U., Kishimoto, A., and Nishizuka, Y. (1989). The protein kinase C family: Heterogeneity and its implications. *Ann. Rev. Biochem.*, **58**, 31–44.

106 Mustelin, T., Coggeshall, K. M., Isakov, N., and Altman, A. (1990). T cell antigen receptor-mediated activation of phospholipase C requires tyrosine phosphorylation. *Science*, **247**, 1584–7.

107 Koretzky, G. A., Picus, J., Thomas, M. L., and Weiss, A. (1990). Tyrosine phosphatase CD45 is essential for coupling T-cell antigen receptor to the phosphatidyl inositol pathway. *Nature*, **346**, 66–8.

108 Windebank, K. P., Abraham, R. T., Powis, G., Olsen, R. A., Barna, T. J., and Leibson, P. J. (1988). Signal transduction during human natural killer cells activation: inositol phosphate generation and regulation by cyclic AMP. *J. Immunol.*, **141**, 3951–7.

109 Edwards, B. S., Nolla, H. A., and Hoffman, R. R. (1989). Relationship between target cell recognition and temporal fluctuations in intracellular CA^{2+} of human NK cells. *J. Immunol.*, **143**, 1058–65.

110 Steele, T. A. and Brahmi, Z. (1988). Phosphatidylinositol metabolism accompanies early activation events in tumor target cell-stimulated human natural killer cells. *Cell Immunol.*, **112**, 402–13.

111 Atkinson, E. A., Gerrard, J. M., Hildes, G. E., and Greenberg, A. H. (1989). Production of inositol-phospholipid-derived second messengers in a rat natural killer cell line exposed to susceptible tumor targets. *Nat. Immun. Cell Growth Regul.*, **8**, 223–30.

112 Paya, C. V., Schoon, R. A., and Leibson, P. J. (1990). Alternative mechanisms of natural killer cell activation during herpes simplex virus infection. *J. Immunol.*, **144**, 4370–75.

113 Cassatella, M. A., Aneg'on, I., Cuturi, M. C., Griskey, P., Trinchieri, G., and Perussia, B. (1989). FcγR (CD16) interaction with ligand induces Ca^{2+} mobilization and phosphoinositide turnover in human natural killer cells. Role of Ca^{2+} in FcγR (CD16)-induced transcription and expression of lymphokine genes. *J. Exp. Med.*, **169**, 549–67.

114 Macintyre, E. A., *et al.* (1989). Binding of monoclonal antibody to CD16 causes calcium mobilization in large granular lymphocytes but inhibits NK killing. *Immunology*, **66**, 459–65.

115 Pantaleo, G., *et al.* (1988). Human cytolytic cell clones lacking surface expression of T cell receptor α/β or γ/δ. Evidence that surface structures other than CD3 or CD2 molecules are required for signal transduction. *J. Exp. Med.*, **168**, 13–24.

116 Leibson, P. J., Midthun, D. E., Windebank, K. P., and Abraham, R. T. (1990). Transmembrane signaling during natural killer cell-mediated cytotoxicity. *J. Immunol.*, **145**, 1498–1504.

117 Inoue, K., Yamamoto, T., and Toyoshima, K. (1990). Specific expression of human c-*fgr* in natural immunity effector cells. *Mol. Cell. Biol.*, **10**, 1789–90.

118 Gutkind, J. S. and Robbins, K. C. (1989). Translocation of the FGR protein-tyrosine kinase as a consequence of neutrophil activation. *Proc. Natl. Acad. Sci. U.S.A.*, **86**, 8783–7.

119 Ullrich, A. and Schlessinger, J. (1990). Signal transduction by receptors with tyrosine kinase activity. *Cell*, **20**, 203–12.

120 Kypta, R. M., Goldberg, Y., Ulug, E. T., and Courtneidge, S. S. (1990). Association between the PDGF receptor and members of the *src* family of tyrosine kinases. *Cell*, **62**, 481–92.

121 June, C. H., *et al.* (1990). Inhibition of tyrosine phosphorylation prevents T-cell receptor-mediated signal transduction. *Proc. Natl. Acad. Sci. U.S.A.*, **87**, 7722–6.

122 June, C. H., Fletcher, M. C., Ledbetter, J. A., and Samelson, L. E. (1990). Increases in tyrosine phosphorylation are detectable before phospholipase C activation after T cell receptor stimulation. *J. Immunol.*, **144**, 1591–9.

123 Downward, J., Graves, J. D., Warne, P. H., Rayter, S., and Cantrell, D. A. (1990). Stimulation of p21ras upon T-cell activation. *Nature*, **346**, 719–23.

124 Imboden, J. B. and Stobo, J. D. (1985). Transmembrane signalling by the T cell antigen receptor complex generates inositol phosphates and releases calcium ions from intracellular stores. *J. Exp. Med.*, **161**, 446–56.

125 Imboden, J. B., Shoback, D. M., Pattison, G., and Stobo, J. D. (1986). Cholera toxin inhibits the T-cell antigen receptor-mediated increases in inositol trisphosphate and cytoplasmic free calcium. *Proc. Natl. Acad. Sci.*, U.S.A., **83**, 5673–7.

126 O'Shea, J. J., Urdahl, K. B., Luong, H. T., Chused, T. M., Samelson, L. E., and Klausner, R. D. (1987). Aluminum fluoride induces phosphatidylinositol turnover, elevation of cytoplasmic free calcium, and phosphorylation of the T cell antigen receptor in murine T cells. *J. Immunol.*, **139**, 3493–9.

127 Mandler, R. ., Seamer, L. C., Whitlinger, D., Lennon, M., Rosenberg, E., and Bankhurst, A. D. (1990). Human natural killer cells express Na$^+$ channels. A pharmacologic flow cytometric study. *J. Immunol.*, **144**, 2365–70.

128 Aneg'on, I., Cuturi, M. C., Trinchieri, G., and Perussia, B. (1988). Interaction of Fc receptor (CD16) ligands induces transcription of interleukin 2 receptor (CD25) and lymphokine genes and expression of their products in human natural killer cells. *J. Exp. Med.*, **167**, 452–72.

129 Harris, D. T., Travis, W. W., and Koren, H. S. (1989). Induction of activation antigens on human natural killer cells mediated through the Fc-γ receptor. *J. Immunol.*, **143**, 2401–6.

130 Schmidt, R. E., *et al.* (1988). T11/CD2 activation of cloned human natural killer cells results in increased conjugate formation and exocytosis of cytolytic granules. *J. Immunol.*, **140**, 991–1002.

131 Mason, A., Bernard, A., Smyth, M. J., and Ortaldo, J. R. (1991).

Role of CD2 in regulation of CD3⁻ LGL function. *Eur. Cytokine Net.*, **2**, 31–7.

132 Argov, S., Hebdon, M., Cuatrecasas, P., and Koren, H. S. (1985). Phorbol ester-induced lymphocyte adherence: selective action on NK cells. *J. Immunol.*, **134**, 2215–22.

133 Young, H. A. and Ortaldo, J. R. (1987). One-signal requirement for interferon-gamma production by human large granular lymphocytes. *J. Immunol.*, **139**, 724–7.

134 Procopio, A. D., *et al.* (1989). Effects of protein kinase C (PK-C) activators and inhibitors on human large granular lymphocytes (LGL): role of PK-C on natural killer (NK) activity. *Cell. Immunol.*, **118**, 470–81.

135 Meyer, T., *et al.* (1989). A derivative of staurosporine (CGP 41 251) shows selectivity for protein kinase C inhibition and *in vitro* antiproliferative as well as *in vivo* antitumor activity. *Int. J. Cancer*, **15**, 851–6.

136 Ito, M., Tanabe, F., Sata, A., Ishida, E., Takami, Y., and Shigeta, S. (1989). Inhibition of natural killer cell-mediated cytotoxicity by ML-9, a selective inhibitor of myosin light chain kinase. *Int. J. Immunopharmacol.*, **11**, 185–90.

137 Steele, T. A. and Brahmi, Z. (1988). Inhibition of human natural killer cell activity by the protein kinase C inhibitor 1-(5-isoquinolinesulfonyl)-2-methylpiperazine is an early but post-binding event. *J. Immunol.*, **141**, 3164–9.

138 Ortaldo, J. R., Young, H. A., and Varesio, L. (1989). Modulation of CD3⁻ LGL functions by agonist and antagonists of PKC: Effects of NK and LAK activity and production of IFNγ. *J. Immunol.*, **143**, 366–71.

139 Hager, C. C., Petroni, K. C., Boyce, M. A., Forester, L. D., and Oeltmann, T. N. (1990). A possible role for protein kinase C activity but not cyclic nucleotide-dependent protein kinases in human natural killer cell lytic activity. *Lymphokine Res.*, **9**, 1–14.

140 Chow, S. C. and Jondal, M. (1990). A central role for phosphoinositide hydrolysis in activating the lytic mechanism of human natural killer cells. *Immunology*, **70**, 106–110.

141 Lyall, R. M., Zilberstein, A., Gazit, A., Gilon, C., Levitzki, A., and Schlessinger, J. (1989). Tyrophostins inhibit epidermal growth factor (EGF)-receptor tyrosine kinase activity in living cells and EGF-stimulated cell proliferation. *J. Biol. Chem.*, **264**, 14503–9.

142 Edwards, B. S., Hawkins, M. J., and Borden, E. C. (1983). Correlation between *in vitro* and systemic effects of native and recombinant interferons-alpha on human natural killer cell cytotoxicity. *J. Biol. Response Mod.*, **2**, 409–17.

143 Giovarelli, M., *et al.* (1988). Obligatory role of IFN-gamma in induc-

tion of lymphokine-activated and T lymphocyte killer activity, but not in boosting of natural cytotoxicity. *J. Immunol.*, **141**, 2831–6.

144 Sayers, T. J., Mason, A. T., and Ortaldo, J. R. (1986). Regulation of human natural killer cell activity by interferon-g: Lack of a role in interleukin 2-mediated augmentation. *J. Immunol.*, **136**, 2176–80.

145 Trinchieri, G., Matsumoto-Kobayashi, M., Clark, S. C., Seehra, J., London, L., and Perussia, B. (1984). Response of resting human peripheral blood natural killer cells to interleukin 2. *J. Exp. Med.*, **160**, 1147–69.

146 Pistoia, V., Zupo, S., Corcione, A., and Ferrarini, M. (1989). Promotion and inhibition of haemopoiesis by NK cells: a model for immune-mediated haemopoietic suppression. *Clin. Exp. Rheumatol. (Italy)*, **7**, (Suppl. 3), S91–4.

147 Djeu, J. Y., Lanza, E., Hopel, A. J., and Ihle, J. N. (1982). Natural cytotoxic activity of mouse spleen cell cultures maintained with interleukin-3. In *NK cells and other natural effector cells* (ed. R. E. Herberman), pp. 917–21. Academic Press, New York.

148 Bloom, E. T. and Babbitt, J. T. (1990). Prostaglandin E2, monocyte adherence and interleukin-1 in the regulation of human natural killer cell activity by monocytes. *Nat. Immun. Cell Growth Regul.*, **9**, 36–48.

149 Shirakawa, F., Tanaka, Y., and Eto, S. (1986). Effect of interleukin 1 on the expression of interleukin 2 receptor (Tac antigen) on human natural killer cells and natural killer-like cell line (YT cells). *J. Immunol.*, **137**, 551–6.

150 Luger, T. A., *et al.* (1989). IFN-β2/IL-6 augments the activity of human natural killer cells. *J. Immunol.*, **143**, 1206–9.

151 Stotter, H., Custer, M. C., Bolton, E. S., Guedez, L., and Lotze, M. T. Interleukin-7 induces human lymphokine activated killer (LAK) cell activity and is regulated by interleukin-4. *J. Immunol.*, **146**, 150–5.

152 Grimm, E. A., Mazumber, A., Zhang, H. Z., and Rosenberg, S. A. (1982). Lymphokine activated killer cell phenomenon. Lysis of natural killer-resistant fresh solid tumor cells by interleukin 2-activated autologous human peripheral blood lymphocytes. *J. Exp. Med.*, **155**, 1823–41.

153 Grimm, E. A., Ramsey, K. M., Mazumder, A., Wilson, D. J., Djeu, J. Y., and Rosenberg, S. A. (1983). Lymphokine-activated killer cell phenomenon. II. Precursor phenotype is serologically distinct from peripheral T lymphocytes, memory cytotoxic thymus-derived lymphocytes, and natural killer cells. *J. Exp. Med.*, **157**, 884–97.

154 Yamada, S., Ruscetti, F. W., Overton, W. R., Herberman, R. B., Birchenall-Sparks, M. C., and Ortaldo, J. R. (1987). Regulation of human large granular lymphocyte and T cell growth and function by recombinant interleukin 2. I. Induction of interleukin 2 receptor and

promotion of growth of cells with enhanced cytotoxicity. *J. Leuc. Biol.*, **41**, 505–17.

155 Kornbluth, J. and Hoover, R. G. (1988). Changes in gene expression associated with IFN-beta and IL-2-induced augmentation of human natural killer cell function. *J. Immunol.*, **141**, 3234–40.

156 Allavena, P. and Ortaldo, J. R. (1984). Characteristics of human NK clones: Target specificity and phenotype. *J. Immunol.*, **132**, 2363–9.

157 Hori, T., Uchiyama, T., Onishi, R., Kamio, M., Umadome, H., and Tamori, S. (1988). Characteristics of the IL-2 receptor expressed on large granular lymphocytes from patients with abnormally expanded large granular lymphocytes. *J. Immunol.*, **140**, 4199–203.

158 Tsudo, M., Goldman, C. K., Bongiovanni, K. F., Chan, W. C., and Winton, E. F. (1987). The p75 peptide is the receptor for interleukin 2 expressed on large granular lymphocytes and is responsible for the interleukin 2 activation of these cells. *Proc. Natl. Acad. Sci. U.S.A.*, **84**, 5394–8.

159 Sharon, M., Siegel, J. P., Tosato, G., Yodoi, J., Gerrard, T. L., and Leonard, W. J. (1988). The human interleukin 2 receptor beta chain (p70). Direct identification, partial purification, and patterns of expression on peripheral blood mononuclear cells. *J. Exp. Med.*, **167**, 1265–70.

140 Caligiuri, M. A., Zmuidzinas, A., Manley, T. J., Levine, H., Smith, K. A., and Ritz, J. (1990). Functional consequences of interleukin 2 receptor expression on resting human lymphocytes. Identification of a novel natural killer cell subset with high affinity receptors. *J. Exp. Med.*, **171**, 1509–26.

161 Kuziel, W. A. and Greene, W. C. (1990). Interleukin-2 and the IL-2 receptor: new insights into structure and function. *J. Invest. Dermatol.*, **94**, 27S–32S.

162 Smyth, M. J., *et al.* (1990). Interleukin 2 induction of pore-forming protein gene expression in human peripheral blood CD8+ T cells. *J. Exp. Med.*, **171**, 1269–1281.

163 Ferris, D. K., Willette-Brown, J., Ortaldo, J. R., and Farrar, W. L. (1989). IL-2 regulation of tyrosine kinase activity is mediated through the p70-75 beta-subunit of the IL-2 receptor. *J. Immunol.*, **143**, 870–6.

164 Saltzman, E. M., Thom, R. R. and Casnellie, J. E. (1988). Activation of a tyrosine protein kinase is an early event in the stimulation of T lymphocytes by interleukin-2. *J. Biol. Chem.*, **263**, 6956–9.

165 Koyasu, S., *et al.* (1987). Interleukin 3-specific tyrosine phosphorylation of Mr 150000 in multifactor-dependent myeloid cell lines. *Embo J.*, **6**, 3979–85.

166 Einspahir, K. J., Abraham, R. T., Dick, C. J., and Leibson, P. J. (1990). Protein tyrosine phosphorylation and p56^{lck} modification in IL-2 or phorbol ester-activated human natural killer cells. *J. Immunol.*, **145**, 1490–7.

167 Sharon, M., Gnarra, J. R., and Leonard, W. J. (1989). The beta-chain of the IL-2 receptor (p70) is tyrosine-phosphorylated on YT and HUT-102B2 cells. *J. Immunol.*, **143**, 2530–3.
168 Asao, H., Takeshita, T., Nakamura, M. Nagata, K., and Sugamura, K. (1990). Interleukin 2 (IL-2)-induced tyrosine phosphorylation of IL-2 receptor p75. *J. Exp. Med.*, **171**, 637–44.
169 Turner, B. C., Rapp, U. R., App, H., Greene, M. I., Dobashi, K., and Reed, J. C. (1991). Interleukin 2 induces tyrosine phosphorylation and activation of p72-74 Raf-1 kinase in a T cell. *Proc. Natl. Acad. Sci. U.S.A.*, **88**, 1227–31.
170 Moreno, S. and Nurse, P. (1990). Substrates for p34^{cdc2}: *In vivo veritas*? *Cell*, **61**, 549–51.
171 Weinberg, R. A. (1990). The retinoblastoma gene and cell growth control. *TIBS*, **15**, 199–201.

2 Natural killer cells in haematopoiesis

G. TRINCHIERI

1   Introduction

In adult animals, blood cells are formed in the central haematopoietic organs, the bone marrow and primary lymphoid organs, and from these organs mature blood cells enter the circulation or home to other tissues. The process of blood cell formation is known as haematopoiesis. Because most blood cells have a limited life span, from a few hours for neutrophilic granulocytes to a few months for erythrocytes, they are continuously regenerated. The human bone marrow, in normal physiological conditions, produces approximately 10^{10} erythrocytes and 4×10^{8} leucocytes every hour. In addition, in conditions of physical stress, such as bacterial infection or anoxia, a rapid increase in the production of cells of a particular lineage is induced. Such a complex system requires a very fine system of regulation in order to be able to maintain the haematopoietic homeostasis and to re-establish it after temporary modifications. A complex series of cellular interactions and humoral factors modulates these regulatory mechanisms. A knowledge of the different levels at which these regulatory mechanisms operate is necessary for any understanding of the possible role in this system of lymphocytes in general and natural killer (NK) cells in particular.

All haematopoietic cells, lymphoid and myeloid, are derived from totipotent immuno-haematopoietic stem cells. These stem cells are mostly non-cycling and characterized by a potent self-renewal ability with limited commitment to differentiation (1). Some of the progeny of the stem cells become committed to differentiate, first to lymphoid or myeloid stem cells, maintaining a multipotentiality to differentiate to different lineages, and then to the progenitor cells of a single lineage. The decision of a stem cell to self-renew or to become committed to differentiate is generally interpreted as a stochastic process of 'birth' (self-renewal) or 'death' (differentiation) (2, 3). The production of the various cell types is regulated at the level of the stem cell population as a whole and not at a level of individual stem cells; control mechanisms act on the probability of self-renewal or commitment (2). The haematopoietic microenvironment most likely does not induce commitment of the stem cells, as originally proposed (4), but is necessary for the continuous differentiation and proliferation of committed progenitor cells. Similarly, humoral factors, such as colony-stimulating factor (CSF), are not able to induce commitment of stem cells but only to regulate the maturation of the cells of different lineages (5).

Stem cells and progenitor cells in the bone marrow are less than 1 in 100 000 cells and 1 in 100 cells, respectively (6). Most studies have identified these cells by their ability to generate differentiated progeny *in vivo* or *in vitro*. Immuno-haematopoietic stem cells are identified by their ability to reconstitute an irradiated animal (7), multipotent myeloid stem cells by the ability to form large colonies of mixed lineage in the spleen of irradiated mice (colony-forming unit-spleen, CFU-S) (8), and committed progenitor cells by the ability to form colonies *in vitro* in semi-solid medium (various types of CFU) (9, 10).

Various types of CFU can be studied *in vitro*. Multipotent progenitors give rise to mixed colonies (CFU-granulocyte, erythrocyte, macrophage, megakaryocyte or CFU-GEMM) (11). A common progenitor cell for neutrophils and macrophages gives rise to dual lineage colonies (CFU-granulocyte, macrophage or CFU-GM). Two types of human CFU-GM have been described: one (day 14 CFU-GM or late appearing CFU-GM) that forms colonies after 14 days of culture and represents a more immature progenitor cell; and one (day 7 CFU-GM or early appearing CFU-GM) that forms colonies after 7 days of culture and represents a more differentiated progenitor cell (12). Mixed colonies are observed predominantly when CSF acting on early cells, such as granulocyte-macrophage-CSF (GM-CSF) or IL-3, are added to the cultures. Lineage-restricted CSF, such as granulocyte-CSF (G-CSF) and macrophage-CSF (M-CSF), mostly determine the growth of single lineage colonies (CFU-G and CFU-M, respectively). Erythroid progenitor cells, similar to CFU-GM, can be divided in two groups; (1) burst forming units-erythroid (BFU-E) are early erythroid progenitors which, in the presence of a burst-promoting activity (BPA, represented by GM-CSF, IL-3, and possibly other CSF) and erythropoietin (Epo), form large irregularly shaped colonies; and (2) CFU-E which are more mature erythroid progenitor cells that, in the presence of Epo alone, rapidly form small colonies in culture. CFU-GEMM, day 14 CFU-GM, and BFU-E are present both in human bone marrow and in peripheral blood, whereas day 7 CFU-GM and CFU-E are present only in the bone marrow (13). Single lineage CFU are also formed by eosinophils (CFU-E), megakaryocytes (CFU-Meg), and basophil/mast cells (CFU-mast).

The bone marrow contains different types of fixed non-haematopoietic cells, which form the marrow stroma, and consist of endothelial cells, reticular epithelial cells, fibroblasts, endosteal cells, and adipocytes. The bone marrow stroma provides the conditions for haematopoiesis to occur, including the production of the various forms of CSF. However, differentiated haematopoietic cells which are present in the bone marrow also play a role in the regulation of haematopoiesis. Various negative and positive feedback mechanisms are mediated by differentiated cells, and macrophages in particular play a central role in the differentiation of different haematopoietic lineages. Lymphocytes represent a significant subset of bone marrow cells and changes can be observed in their number or activation state in certain pathological conditions. Both CD4$^+$ and CD8$^+$ subsets of T-cells are present, whereas mature NK cells are rare, although this cell type originates in the bone marrow. A role for T- and NK cells, in regulating haematopoiesis is suggested by pathological situations where alteration of lymphocyte activity and distribution results in severe alteration of haematopoiesis, usually depression, at the level of one or several lineages. In a classical study, Bagby and coworkers (14) analysed over 200 patients with neutropenia of different etiologies, and found that in a fifth of them, removal of lymphocytes from bone marrow cells resulted in significantly increased formation of CFU-GM colonies, suggesting that lymphocyte-mediated suppression of granulopoiesis takes place in these patients. In this chapter, the evidence suggesting a role for NK cells in the regulation of haematopoiesis will be discussed.

2   Evidence for the role of NK cells in the regulation of haematopoiesis: murine experiments *in vivo*

2.1   Hybrid resistance

The first indication of a role for NK cells in haematopoiesis came from the studies of Cudkowicz and co-workers (15–18) into the phenomenon of hybrid resistance to parental bone marrow transplantation. Parental haematopoietic tumours or bone marrow grafts do not survive in lethally irradiated F1 mice, even if these animals accept grafts of skin or any other type of parental tissue. These findings contrast with classical transplantation studies showing the existence of codominant histocompatibility (H) genes. Hybrid resistance has been explained assuming the existence of cellular determinants, expressed only on haematopoietic cells, and controlled by a class of non-codominant genes, the haemtopoietic histocompatibility (Hh) genes. The major Hh locus, Hh-1, has been mapped by classical genetic studies in the murine MHC, between H-2S and the centromeric side of H-2D (19, 20), although studies with one H-2D transgenic mouse strain has suggested the possibility that the H-2D locus itself is responsible for the graft rejection (21).

The characteristics of the effector cells mediating hybrid resistance has identified them as NK cells (18). These include genetic control in different mouse strains, radioresistance, age of maturation, bone marrow dependence, thymus independence, lack of immunological memory, and sensitivity to split-dose irradiation. Much more compelling evidence for NK cells being the effector cells in marrow graft rejection was provided by experiments in which NK cells were depleted *in vivo* using anti-NK cell antibodies (22, 23). The ability of T-cell clones with NK-like cytotoxic activity (24) or of a CD3[+] T-cell subset expressing the NK marker NK1.1 (25), to transfer the ability to reject parental marrow graft has suggested that the effector cells of hybrid resistance are not NK cells but belong to a T-cell subset sharing surface markers with NK cells. However, the presence of hybrid resistance in Scid mice, which do not have functional

T- or B-cells, mitigates against T-cells being the major effector cell type (26).

The similarity between hybrid resistance effector cells and NK cells excludes the most pertinent property of hybrid resistance, that is, its immunogenetic specificity. A possible role for natural antibodies in directing the specificity of NK cells, with an *in vivo* mechanism similar to antibody-dependent cell-mediated cytotoxicity (ADCC), was proposed on the basis of serum transfer experiments (27). However, natural antibodies were found only in some strains and serum was shown to transfer allogeneic resistance but not hybrid resistance (27). The ability of Scid mice, which have neither mature B-cells nor antibodies, to reject parental grafts, definitively excludes the possibility that antibody-directed mechanisms play a major role in hybrid resistance (26). The ability of NK cells, known for their non-specific non-MHC restricted cytotoxicity, in the absence of antigen-specific receptors, to display *in vivo* a fine immunogenetically controlled specificity remains, therefore, an enigma.

The hybrid resistance phenomenon has been difficult to reproduce *in vitro*. In one study (28), purified murine NK cells from F1 mice have been shown to suppress parental CFU-GM colony formation more efficiently than syngeneic colony formation, but this complex experimental system was not amenable to a detailed genetic study. Recently, Ciccone and co-workers (29, 30) showed that certain human CD3⁻ NK cell clones generated from responder cells in an allogeneic mixed leucocyte culture (MLC) are able to lyse specifically the stimulator cells used in the MLC. The structures recognized by these allogeneic clones have been shown to be controlled by non-codominant genes situated in the MHC region of human chromosome 6. The genetic control of these specificities closely resemble that of Hh-1. These findings, which are still unexplained in terms of NK receptors involved in allogeneic recognition and of generation of their specificity, open new and fascinating prospects in the study of NK cell activity.

Overall, these and other observations in the field of hybrid resistance show that NK cells have the ability to reject bone marrow grafts. Although most of the hybrid resistance studies have involved measuring formation of CFU-S colonies in the spleen by morphological analysis or isotope incorporation, the ability of NK cells to prevent long-term haematopoietic reconstitution by parental marrow has also been shown (31). The importance of these findings for engraftment failure in clinical transplantation should, therefore, be considered. It is of interest to note that similar to the increased engraftment failure observed in clinical transplantation, elimination of T-cells from murine bone marrow grafts leads to an increased susceptibility to rejection by NK cells (32).

These findings clearly show that NK cells can be active *in vivo* against haematopoietic progenitor cells. Although graft rejection requires allogeneic Hh differences, the observed activity might be a pathological amplification of the physiological role of NK cells in the syngeneic environment. On the basis of these observations, Cudkowicz and Hochman in 1979 (17) originally postulated the hypothesis that NK cells might 'engage in regulated reactions against self to ensure homeostasis'.

2.2   Experimentally induced haematopoietic disorders involving viruses

During viral infection, a NK cell response, which usually peaks at 3 days post-infection, is followed by a cytotoxic lymphocyte (CTL) response (33, 34), suggesting that NK cells, together with interferon (IFN) and other natural resistance mechanisms, represent the first line of defence of the organism against infection. A particularly impressive NK response is induced in mice acutely infected with lymphocyte choriomeningitis virus (LCMV). In these mice, high levels of IFN and an increased number and activity of NK cells in spleen, peritoneum, liver, lung, bone marrow, and peripheral blood are observed (35). The increased number of NK cells, which have activated blast morphology and are proliferative, is due to increased production in the bone marrow (35).

Although adult mice injected intraperitoneally with LCMV undergo a relatively mild disease, this is followed by marked immunological and haematological depression. In the first week of infection, there is a profound depression in the number of CFU-S and CFU-GM, and in the level of erythropoiesis as measured by ^{51}Fe uptake; on day 10, the number of CFU-S and erythropoiesis returns to normal values in the spleen, but a bone marrow depression is measurable for 3 weeks (36, 37). NK cells with activated morphology are present in the bone marrow of these mice; these NK cells are probably relatively immature cells that are retained in the bone marrow, suggesting that infection may increase the transit time of differentiated NK cells in the bone marrow (38). Using adoptive transfer protocols or treatment of the animals with anti-asialo GM1 antiserum, it was demonstrated that NK cells are directly responsible for the haematopoietic depression observed in the first phase of the disease (38).

The studies with LCMV-infected mice, or with adoptively transferred cells from these animals, demonstrate that *in vivo* activated NK cells can suppress growth and proliferation of syngeneic haematopoietic progenitor cells and that this inhibitory activity is observed not only in the spleen or peripheral blood, that is, tissues with a high NK cell number, but also in the bone marrow, an organ that normally presents low NK cell cytotoxic activity and number.

Many types of viral infection can induce transient or permanent suppression of one or more haematopoietic lineages. Some of these effects may be mediated by NK or T effector cells, although others may depend on a direct cytopathic effect of the virus on the progenitor cells or on an induction by the virus of haematopoiesis inhibitors, such as IFN (39, 40).

2.3 Effects of depletion of NK cells *in vivo* on haematopoiesis

Animals with deficient NK cell activity, such as Beige mice, apparently have normal haematopoiesis. However, this observation does not exclude a role for NK cells in haematopoietic homeostasis, since NK cells in these mice are present, but just defective in their cytotoxicity activity. Also, because haematopoiesis is regulated by different and probably redundant mechanisms, in the case of a chronic defect of NK cells it should be expected that alternative mechanisms intervene to maintain the homeostatic equilibrium. The best demonstration of this ability of the haematopoietic system to regulate itself even in the presence of severe defects is offered by the W/W^v mice. These mice, due to the deficient activity of the stem cell factor receptor/c-*kit* (41), have a severe depression in number of CFU-S, but an almost normal number of differentiated haematopoietic cells (42, 43).

The availability of monoclonal antibodies anti-NK1.1 with acceptable specificity for NK cells has allowed various investigators to address the question directly as to which haematological changes are induced by acute depletion of NK cells. Hansson and co-workers (44) have shown that chronic treatment with anti-NK1.1 serum induces a several-fold increase in the number of CFU-GM in the spleen, but not in the bone marrow. Pantel and Nakeff (45) showed that in the anti-NK-treated animals CFU-Meg in the bone marrow were decreased by 60 per cent in number and their proliferation completely abolished. On the contrary, the proportion of cycling cells in the BFU-E was increased two-fold, although their number was not affected. Pantel and others (46) analysed the recovery from radiation induced myelosuppression in mice depletion of NK cells. Anti-NK-treated mice showed a significantly faster recovery of CFU-S and CFU-GM, whereas recovery of CFU-Meg, and possibly CFU-E, was slower than in control mice.

Overall, experimental evidence in mice has demonstrated that NK cells in physiological conditions have a regulatory effect on haematopoiesis, with a stimulatory effect on megakaryocytosis and erythropoiesis, but an inhibitory one on phagocytopoiesis and on myeloid stem cells. These effects can be demonstrated both in the spleen and in the bone marrow, although differences in the findings between the two organs has been observed. The role of NK cells in the regulation of haematopoiesis is much more dramatic in the case of pathological activation by viruses or IFN. In these instances, a significant inhibitory effect at the bone marrow level which results in anaemia and neutropenia can be observed. Finally, when confronted with allogeneic haematopoietic cells expressing incompatible

patible Hh-1 specificities, NK cells are particularly effective and can prevent haematopoietic reconstitution completely.

3 Evidence for the role of NK cells in the regulation of human haematopoiesis

The evidence for a physiological role of NK cells in the regulation of haematopoiesis in humans is indirect and based on extrapolation from *in vivo* findings in patients with haematopoietic abnormalities or from *in vitro* studies. One of the major difficulties in identifying a role for NK cells in clinical haematopoietic disorders is the imprecise methods currently used to identify NK cells. The LGL morphology of NK cells is relatively specific for this cell type in human peripheral blood from healthy donors, but can be expressed by activated T-cells. In many studies of NK cells and haematopoiesis, the anti-Leu7/HNK-1 antibody (CD57) was used for identification of NK cells: CD57 antigen, however, is not only expressed on a proportion of NK cells, but also on a subset of CD8$^+$ T-cells and on some CD4$^+$ T-cells. Interpretation of the results observed with anti-Leu7 antibody was made even more difficult by the facts that the CD57 antigen is expressed on activated T-cells found in many patients and that most CD57$^+$ cells have a LGL morphology. The transmembrane form of low affinity receptor for Fc of IgC (FcγRIIIA or CD16 antigen) is expressed virtually on all peripheral blood NK cells, but also in a rare CD16$^+$ CD3$^+$ subset of T-cells, which can be expanded in pathological situations, such as in most patients with LGL lymphocytosis. The CD16 antigen is also frequently absent from NK cells found in association with tissues. The CD56 antigen (NCAM, Leu19) is expressed on almost all NK cells, but it is also expressed on a minor subset of T-cells and, upon blast formation and activation, can be found on almost all T-cells with non-MHC-restricted NK-like cytotoxic activity. In order to identify unambiguously NK cells in clinical samples, it is necessary to use several criteria, including expression of CD56, CD16, and CD2 antigens, lack of surface expression of T-cell receptor (TCR) and of TCR-associated CD3 complex, and lack of rearrangement of the TCR genes.

In humans, bone marrow depression is often associated with the presence of activated CD3$^+$ and CD8$^+$ large granular T-cells which often express CD57 antigen and are able to inhibit *in vitro* haematopoietic colony formation. These cells, which have also been identified in normal donors, are often able to suppress haematopoietic progenitor cells in a MHC Class II restricted fashion, a surprising finding with CD8$^+$ T-cells. CD3$^+$, CD8$^+$, and CD57$^+$ T-cells with colony-inhibiting activity have also been found in patients rejecting T-cell-depleted allogeneic bone marrow transplants and shown to have a MHC Class II restricted specificity (47, 48). Contrasting with the finding in patients achieving engraftment, an early recovery of NK cells and NK cytotoxic activity was consistently absent from the blood of the patients rejecting the transplanted bone marrow (47).

Because activated T-cells found in patients can express NK cell markers and morphology, many of the early studies that have identified haemato-poietic-inhibiting cells with NK cell markers in patients may no longer be considered as providing sufficient evidence for a role of NK cells in the haematopoietic dysfunction. A role for suppressor lymphocytes was demonstrated in red cell aplasia during B-cell chronic lymphocytic leukaemia, in the neutropenia and transient bone marrow hypoplasia found in approximately 10 per cent of the patients with Epstein–Barr virus (EBV)-induced mononucleosis, and in severe acquired aplastic anaemia. This latter disease is a heterogeneous group of disorders of various etiologies of which lymphocyte-mediated suppression may represent only one of the pathogenetic mechanisms. Bone marrow from patients with aplastic anaemia contain cells with much higher cytotoxic activity against NK-sensitive target cells than the bone marrow from healthy donors (49). Removal of bone marrow lymphocytes from aplastic patients resulted in increased haematopoietic colony growth *in vitro* in only 10 per cent of patients (14, 50), although *in vivo* treatment with antilymphocyte globulins induces at least a partial improvement in about half of the patients (51, 52). As with other diseases, there is no definitive evidence that the suppressor cells in aplastic anaemia are NK cells. However, at least one case of rejection of bone marrow from an identical twin, was attributed to cells with NK characteristics in the host aplastic anaemia patient (53). Patients with Felty's syndrome (neutropenia, arthritis, splenomegaly) and those with adult-onset cyclic neutropenia have a moderate expansion of CD57$^+$ LGL. However, in most of the Felty's syndrome patients the CD57$^+$ LGLs are CD3$^+$ T-cells (54), whereas in some cyclic neutropenia patients, expression of LGLs with a phenotype characteristic of NK cells was described (55).

Both acute and chronic monoclonal expansions of NK cells have been observed. In one of the very rare cases of acute NK cell leukaemia, the patient had CD2$^+$, CD3$^-$, CD16$^-$, and CD56$^+$ malignant cells with suppressor activity on autologous CFU-GM and CFU-E colony formation; the patient died with severe neutropenia and sepsis two months after the diagnosis (56). Patients with LGL lymphocytosis are usually characterized by a very chronic progression of the disease, often associated with neutropenia or anaemia. About 90 per cent of the patients with LGL lymphocytosis display a monoclonal expansion of CD3$^+$, CD8$^+$, and CD56$^+$ T-cells. In 10 per cent of patients, however, the LGLs are typical NK cells with a CD3$^-$, CD16$^+$, and CD56$^+$ phenotype. In this latter group of patients, neutropenia and anaemia are also frequently observed and LGL/NK cells of these patients suppress proliferation/differentiation of haematopoietic progenitor cells *in vitro* (57).

Overall, these clinical studies do not provide strong evidence in favour of a major role for NK cells in regulating human haematopoiesis, although some studies clearly demonstrate that NK cells *in vivo* can suppress autologous haematopoietic progenitor cells. If one considers the pathological findings an exaggeration of the physiological role of NK cells, then the clinical data are consistent with a role of NK cells in haematopoietic homeostasis. The major reason for the lack of clinical information on the role of NK cells in haematopoietic pathology is that the technology used in most studies was not sufficiently accurate to distinguish unambiguously between effects mediated by NK cells and those mediated by T-cells. However, most studies seem to suggest that haematopoiesis-suppressive T-cells might be more frequently involved in bone marrow dysfunction or in bone marrow graft rejection than NK cells.

4   Effects of NK cells on haematopoietic progenitor cells: experiments *in vitro*

4.1   Cell-mediated effects

One physiological role which has been suggested for NK cells is that they are involved in the surveillance of primitive cell types and participate in the surveillance against tumours by destroying malignant cells expressing embryonal-type antigenic structures (58). Normal primitive cell types that are sensitive to NK cell-mediated cytotoxicity have been found in the thymus and bone marrow, two organs in which the physiological level of NK cells is low (59–61). However, a direct cytotoxic effect of NK cells on progenitor cells was difficult to demonstrate because the number of the latter is low in bone marrow. Furthermore, the effects of NK cells on haematopoietic progenitor cells may not be limited to cell-mediated cyto-toxicity, but may be dependent on other functions of NK cells, e.g. on cytokine production. Most *in vitro* evidence for a role of NK cells in haematopoiesis comes from experiments in which the ability of NK cells to affect colony formation by progenitor cells was evaluated. These experiments cannot distinguish whether NK cells act directly on the progenitor cells with cytotoxic or cytostatic mechanisms or affect the process of cellular proliferation and differentiation that takes place during the 7 to 14 days required for colony formation *in vitro*. The effects mediated by NK cells in these assays can require cellular contact or be mediated by secreted humoral factors. It is also possible that accessory cells are required for the effect of NK cells: even when 'pure' preparations of NK cells and progenitor cells are used; it is possible that some of the differentiated progeny of the progenitor cells act as accessory cells for the NK cell-mediated effects.

Several early studies *in vitro* showed that cells with NK characteristics [e.g. sheep erythrocyte-rosetting cells expressing FcγR (62) or non-B-, non-T-lymphocytes (63)] were able to inhibit CFU-GM. Hansson and co-workers (64) identified the CFU-GM suppressor cells as NK cells using Percoll gradient separation. They showed that (i) NK cells inhibit both autologous and allogeneic CFU-GM, (ii) the inhibition was enhanced by pretreatment of NK cells with IFN, and (iii) NK-sensitive target cells competed for the inhibition. The inhibitory effect was observed when NK cells and bone marrow cells were incubated together for several hours before plating, suggesting a requirement for cell contact. Several other studies confirmed an inhibitory effect of NK cells on bone marrow CFU-GM and CFU-E (65–67) using semi-purified preparation of NK cells. Degliantoni and co-workers (68, 69) characterized the phenotype of the peripheral blood lymphocytes able to spontaneously suppress haematopoietic colony formation. Those were shown to have the exact phenotype of NK cells, i.e CD16$^+$, CD56$^+$, CD3$^-$, CD5$^-$, CD4$^-$, HLA-DR$^-$, mostly CD2$^+$, and, in part, CD8$^+$ and CD57$^+$. These purified NK cells were shown to inhibit day 14 CFU-GM, CFU-E, and CFU-GEMM colonies, but neither BFU-E nor day 7 CFU-GM (68, 69), contrasting with studies from other authors who showed a preferential inhibition of day 7 CFU-GM (64, 66). IFN-treatment potentiated the ability of purified NK cells to suppress colony formation. The ability of human NK cells to suppress progenitor cells *in vitro* was confirmed by the demonstration that CD3$^-$, IL-2-dependent NK cell clones inhibited progenitor cells when cell contact with bone marrow cells was allowed to take place during a several-hour incubation in liquid cultures (70).

Although the vast majority of laboratories working in this area have reported the ability of NK cells to suppress colony formation from bone marrow progenitor cells (62–69, 71–73), some recent reports have shown no detectable effect of NK cells (74–76). Niemeyer and co-workers (74) did not detect any effect of CD16$^+$ NK cells on enriched bone marrow precursor cells. In this report the only preparation of NK cells used was CD16$^+$ peripheral blood lymphocytes separated by fluorescence-activated cell sorting after staining with the IgM antibody Leu11b. NK cell cytotoxic activity can be down-modulated by cross-linking of the CD16 FcγR (77) and soluble factors, both inhibitory (e.g. TNF) or stimulatory (e.g. GM-CSF), for haematopoiesis can be produced by NK cells subjected to this treatment (78, 79). The results reported by Niemeyer and co-workers (74) are therefore difficult to interpret with any degree of certainty. However, the major difference between their studies and those of others is the use of preparations of progenitor cells more extensively purified than those used by other authors. The presence of contaminant bone marrow cell types could be required for activation of NK cells or to act as accessory cells for the activity of soluble factors, as discussed in more detail in the following section. Two other studies (75, 76), in which the inhibitory effect of NK cells was not observed, used either CD3⁻, IL-2-dependent NK cell clones or IL-2-activated NK cells (LAK) as effector cells. Because, as discussed below, NK cells can produce different soluble factors with both stimulatory or inhibitory effects on haematopoiesis, these negative results may reflect the differential ability of the *in vitro*-grown NK cells to produce either type of factor. CD3⁻ NK cell clones with the ability to suppress colony formation have been described in another study (70) and their activity was attributed in part to production of IFNγ. It would be of interest to compare the ability of inhibitory and non-inhibitory NK cell clones to produce various soluble factors.

The possibility that bone marrow preparations contain a cell type different from progenitor cells which is required for the inhibitory effect of NK cells is also suggested by two studies which indicate that NK cell preparations which exert a strong inhibitory effect on bone marrow progenitor cells have either an inhibitory or only a mild stimulatory effect on the formation of colonies from peripheral blood progenitor cells (66, 73).

In some experimental conditions, NK cells added to haematopoietic progenitor cell cultures induce or enhance colony growth. As mentioned above, purified NK cells enhance growth of CFU-GM colony formed by peripheral blood progenitor cells (66). Also, purified NK cells have a powerful enhancing effect on the formation of CFU-Meg colony formation from human bone marrow (80). Most of the stimulatory effects of NK on haematopoiesis are probably mediated through release of soluble factors, as discussed in the next section.

The majority of studies on the effect of human NK cells on bone marrow progenitor cells demonstrate an inhibitory effect on CFU-GEMM, CFU-GM, and CFU-E colony formation, whereas a stimulatory effect was demonstrated for CFU-Meg. The inhibitory effect of NK cells, however, was not demonstrated on peripheral blood progenitor cells and colony formation by peripheral blood CFU-GM was enhanced by NK cells. The different effects on bone marrow and peripheral blood progenitor cells, when taken together with the difficulty encountered in demonstrating NK cell inhibitory activity in some studies in which marrow progenitor cells were purified or in which NK cells were cultured *in vitro* before being tested, suggest that the inhibitory effect of NK is not due to a simple, one-to-one interaction between an NK cell and a progenitor cell, with a cytotoxic or cytostatic effect. Rather this may involve accessory cells and the production of soluble factors.

4.2   Production by NK cells of soluble factors affecting haematopoiesis

Because of the difficulty of obtaining pure preparations of NK cells before well characterized monoclonal antibodies became available, many early studies which suggested the production of various cytokines by NK cells are now difficult to interpret. More recent studies with purified NK cell preparations have confirmed the ability of NK cells to produce cytokines. Production of IL-1, IFNγ, and B-cell stimulatory factors by NK cells is supported by data from various laboratories (81–84). BPA production by NK cells has been reported (85, 86), and recent data show that this activity can probably be attributed to GM-CSF and/or IL-3 produced by NK cells (87). Production by NK cells of a CSF activity acting on CFU-GM was also suggested in recent studies (66, 88).

Degliantoni and co-workers (68, 69) showed that purified NK cells produce a colony-inhibitory activity (NK-CIA) when co-cultured for several hours with NK-sensitive target cells (such as K562 cells), or with allogeneic or autologous bone marrow cells but not with NK-insensitive cells (such as Raji cells). The inhibitory activity present in the supernatant fluid of the co-cultures and the specificity of the inhibition for different types of CFU was consistent with the hypothesis that the inhibitory effect of NK cells was mediated by NK-CIA and that cell contact between NK cells and bone marrow cells was required for induction of NK-CIA production. These results have been reproduced in studies from several groups (72, 73, 89). HLA-Dr$^+$ bone marrow cells, highly enriched for haematopoietic progenitor cells, induce NK-CIA production, whereas HLA-Dr$^-$ cells depleted of precursor cells, fail to do so, suggesting the possibility that progenitor cells themselves are the inducers of NK-CIA production (68). Nagler and Greenberg (72) extended these findings by identifying the inducer cells in bone marrow as HLA-Dr$^+$ CD34$^+$ cells, a cellular population that comprises all the progenitor cells in the bone marrow. However, purified progenitor cell preparations from peripheral blood are not inhibited by NK cells and do not induce NK-CIA production (73). The CD34$^+$ subset in bone marrow contains cells that are more mature progenitor cells than the CD34$^+$ subset in peripheral blood. These experiments are, therefore, compatible with the hypothesis that the NK-CIA inducer cell is a bone marrow CD34$^+$ progenitor cell at a stage of differentiation more advanced than the early progenitor cells in peripheral blood. However, these results need to be confirmed in order to exclude possible artefacts (e.g. reactivity of anti-CD34 with a cell type different from bone marrow progenitor cells or activation of NK cells via FcγR triggering by the monoclonal antibodies used in the cell separation protocols). The possibility that a bone marrow cell type distinct from progenitor cells is required for induction of NK cell production of NK-CIA also cannot be excluded at the present time.

The inhibitory activity of NK-CIA was shown to be due almost exclusively to tumour necrosis factor alpha (TNFα) produced by NK cells (69, 89). Antibodies against this cytokine completely suppressed NK-CIA (69, 89) and recombinant TNFα, at the same concentration as the TNFα activity present in the NK-CIA-containing supernatant fluids, closely mimicked the inhibitory effect of NK-CIA on progenitor cells (69, 90). However, TNFα is a potent inhibitor of BFU-E (90), whereas NK cells and NK-CIA failed to elicit this effect (68), suggesting that BPA activity produced by NK cells may mask the NK-CIA-mediated inhibition of this type of colony (86, 89). Both NK-CIA and TNFα synergize with IFNγ in inhibiting various types of CFU (69, 90). For example, day 7 CFU-GM colony formation induced by GM-CSF is not inhibited by TNFα/NK-CIA or IFNγ separately, but is markedly suppressed by a combination of both factors at low concentrations (69, 90). Bone marrow or target cell stimulation of freshly isolated peripheral blood NK cells induces the production of a low level of TNFα but not IFNγ (69). However, activated NK cells or IL-2-stimulated NK cells (78, 82) can be induced to produce both IFNγ and TNFα, and a synergistic effect of the two factors on haematopoietic colonies can be expected in such conditions. Similarly, mitogen activated T-cells have been shown to inhibit CFU-GM colonies by the synergistic effect of IFNγ and lymphotoxin, a cytokine which is produced by T-cells and has biological activity and receptor specificity which is virtually identical to TNFα (91).

Most of the inhibitory effects of NK cells on haematopoiesis *in vitro* appear to be mediated by TNFα. However, the effects of TNFα itself on haematopoiesis are extremely complex [for a review see (92)]. TNFα was originally described to be mostly inhibitory on haematopoietic colony formation (69, 90, 93). Studies with purified CD34+ progenitor cells have shown, however, that TNFα increases the proliferation of these cells in response to IL-3 or GM-CSF (94). On more mature progenitor cells, formation of CFU-G and CFU-E colonies is inhibited by TNFα, whereas CFU-M colony formation is enhanced (90, 94). This enhancement of macrophage differentiation is consistent with the ability of TNFα to induce human myeloid cell lines to differentiate to monocyte-macrophages (95). The inhibitory effect of TNFα may be indirect and require accessory cells. TNFα inhibits CFU-E colony formation from bone marrow progenitor cells, but not from progenitor cells derived *in vitro* from BFU-E (96). These results have been explained by the demonstration that TNFα inhibition requires the participation of a soya bean agglutinin-binding accessory cell type present in the bone marrow (96). Because inhibition by TNFα is observed in cultures of purified CD34+ cells (94), it is possible that the accessory cells required for TNFα action are haematopoietic cells derived *in vitro* from the CD34+ progenitor cells. The effects of TNFα *in vivo* are even more complex, ranging from alteration in the circulation of neutrophils and lymphocytes and inhibition of erythropoiesis to protective effects on bone marrow with accelerated restoration of haematopoiesis in animals compromised by sub-lethal doses of cytotoxic drugs or irradiation. This latter protective effect of TNFα (97, 98) is probably due to the proliferative effect on early progenitor cells (94), but also to the ability of TNFα to induce endothelial cells and fibroblasts to produce colony-stimulating factors, such as IL-6, GM-CSF, M-CSF, and G-CSF (99–101).

The production of cytokines by purified NK cells has been studied following stimulation with FcγR ligands (immune complexes or anti-CD16 antibodies) or IL-2 (78, 102). The two stimuli separately induce IFNγ and TNFα production and, when used together, synergistically induce cytokine production (78). Both IL-2 and FcγR ligands induce an increase in the transcriptional rate of the TNFα and IFNγ. Although a synergistic effect on mRNA accumulation is observed with the two stimuli combined, no additional increase in transcription is observed, suggesting that post-transcriptional mechanisms are involved in the synergistic effect of the two stimuli (78). Since stimulation through FcγR activates the lytic mechanism of NK cells in antibody-dependent cell-mediated cytotoxicity, this type of stimulation may mimic NK cell-target interaction. Indeed, IFNγ and TNFα production are observed when cultured NK cells are exposed to target cells and this response is augmented by IL-2 (G. Trinchieri, unpublished observations). NK cells stimulated by FcγR ligands and IL-2 also produce significant levels of GM-CSF and, to a lesser extent, M-CSF (87). Stimulation of purified NK cells with phorbol diesters and a Ca^{2+} ionophore induces the production of IL-3, which is not oberved in NK cells stimulated by FcγR ligands and/or IL-2 (79). These results suggest that NK cells can be easily induced to produce TNFα, IFNγ, GM-CSF, and possibly, M-CSF, whereas production of IL-3 is observed only in conditions of maximal stimulation of the cells. Pistoia and co-workers (89) have presented data suggesting that resting NK cells constitutively produce low levels of GM-CSF, but that this activity is masked upon stimulation of NK cells by the production of TNFα, which antagonizes GM-CSF activity.

The ability of NK cells to enhance the growth of CFU-Meg was also shown to be mediated by soluble factors (80). Several factors with thrombopoietic activity, including IL-3, erythropoietin, and IL-11, have been identified, and it is difficult to determine whether the activity produced by NK cells can be accounted for by one of the known cytokines, or whether it represents a novel factor.

4.3   Inhibitory effects of NK cells on leukaemic progenitor cells

A possible role of NK cells in inhibiting clonogenic growth of leukaemia cells has been suggested (103). Allogeneic NK cells were shown to suppress colony formation from the malignant blasts of three patients with acute leukaemia. Studies in which LAK cells or NK cell clones were incubated with bone marrow preparations containing leukaemic cells demonstrated efficient elimination of the malignant progenitor cells, but not their normal counterparts (75, 76). Studies in rats have shown that bone marrow treated with LAK cells can be purged from leukaemic cells without affecting the ability of the marrow to haematologically reconstitute irradiated recipients (104). These studies suggest that totipotent immuno-haematopoietic stem cells are not destroyed by NK/LAK cells, and that this type of treatment could be considered for purging leukaemic or malignant cells from the bone marrow preparations to be used in clinical autologous bone marrow transplantation.

Donor NK cells recover very early after successful bone marrow transplants and are usually considered to be responsible, at least in part, for the graft versus leukaemia reaction presumed to take place in patients. Since these rapidly expanding NK cells do not seem to have a deleterious effect on the engraftment process, procedures capable of augmenting their production and activation might increase the antileukaemia response in such patients.

5 Summary

Many aspects of the possible effects of NK cells on haematopoiesis are still obscure and most of our knowledge is extrapolated from *in vitro* experiments or from observations in pathological conditions *in vivo*. The regulatory mechanisms used by the haematopoietic system are redundant and it is difficult to demonstrate the essential role of a specific cell type or factor. From the data presented in this review, it is clear, however, that NK cells have some role in maintaining haematopoietic homeostasis, although this cannot be simply accounted for by recognition and subsequent lysis of progenitor cells with which NK cells come into physical contact. All or most of the effects of NK cells on haematopoiesis are probably mediated by non-cytotoxic mechanisms involving primarily cytokine production by these cells, either constitutively or upon activation. NK cells produce a variety of cytokines, some with a prevalently stimulatory effect on haematopoiesis, and some mostly inhibitory. Some cytokines produced by NK cells also have different and sometimes contrasting effects *in vivo* depending on the concentration reached in the haematopoietic tissue and on the haematopoietic status of the organism. For example, TNFα has a suppressive effect on bone marrow haematopoiesis of intact animals, but enhances haematopoietic reconstitution of irradiated animals. It is, therefore, not surprising that, in different experimental (or pathological) conditions *in vivo* and *in vitro*, apparently contrasting results are obtained when the effects of NK cells on haematopoiesis are analysed.

A further level of complexity in the effect of NK cells on haematopoiesis is offered by the requirement of additional cell types that appear to be involved both as stimulator and/or accessory cells for cytokine production by NK cells, and as accessory cells for the inhibitory and stimulatory effects of factors such as TNFα. Some of these accessory functions might be provided by the progenitor cells themselves or by their differentiated progeny, others might be provided by non-haematopoietic cells, such as those making up the bone marrow stroma.

The specificity of the NK cell recognition of allogeneic and syngeneic haematopoietic progenitor cells, the mechanisms of induction of cytokine production, and the characterization of the identity and role of accessory cells are the major unresolved tasks in the study of the role of NK cells in haematopoiesis. The knowledge of these processes will allow us to understand better the role of NK cells *in vivo* in physiological and pathological conditions. Because of the therapeutic use of NK cells or NK cell-activating agents, this knowledge will also have a clinical relevance to the possible therapeutic effects of these cells against malignant haematopoietic cells and the undesirable side-effects on normal haematopoiesis.

References

1 Hogson, G. S. and Bradley, T. R. (1979). Properties of haematopoietic stem cells surviving 5-fluorouracil treatment: evidence for a pre-CFU-S cell? *Nature*, **281**, 381–4.

2 Till, J. E., McCulloch, E. A., and Siminoutch, L. (1964). A stochastic model of stem cell proliferation, based on the growth of spleen colony-forming cells. *Proc. Natl. Acad. Sci. U.S.A.*, **51**, 29–36.

3 Ogawa, M., Porter, P. N., and Nakahata, T. (1983). Renewal and commitment of hemopoietic stem cells (an interpretative review). *Blood*, **61**, 823–9.

4 Trentin, J. J. (1970). Influence of haematopoietic organstroma (haematopoietic inductive microenvironment) on stem cell differentiation. In *Regulation of hematopoiesis* (ed. A. S. Gordon), pp. 161–6. Appleton-Century-Crofts, New York.

5 Ogawa, M. (1987). Stem cell functions assessed in clonal culture. In *Cell physiology of blood*, (ed. R. B. Gunn and J. C. Parker), pp. 39–46. Rockefeller University Press, New York.

6 Harrison, D. E., Stone, M., and Astle, C. M. (1990). Effects of transplantation on the primitive immunohematopoietic stem cell. *J. Exp. Med.*, **172**, 431–7.

7 Harrison, D. E., Astle, C. M., and Lerner, C. (1988). Number and continuous proliferative pattern of transplanted primitive immunohematopoietic stem cells. *Proc. Natl. Acad. Sci. U.S.A.*, **85**, 822–6.

8 Till, J. E. and McCulloch, E. A. (1961). A direct measurement of the radiation sensitivity of normal mouse bone marrow cells. *Radiat. Res.*, **14**, 213–22.

9 Pluznik, D. H. and Sach, L. (1965). The cloning of normal 'mast' cells in tissue culture. *J. Cell. Physiol.*, **66**, 319–24.

10 Bradley, T. R. and Metcalf, D. (1966). The growth of mouse bone marrow cells *in vitro*. *Austr. J. Exp. Biol. Med.*, **44**, 287–99.

11 Fauser, A. A. and Messner, H. A. (1978). Granuloerythropoietic colonies in human bone marrow, peripheral blood, and cord blood. *Blood, 52,* 1243–8.

12 Ferrero, D. *et al.* (1983). Antigenically distinct subpopulations of myeloid progenitor cells (CFU-GM) in human peripheral blood and marrow. *Proc. Natl. Acad. Sci. U.S.A.*, **80**, 4114–18.

13 McCredie, K. B., Hersh, E. M., and Freireich, E. J. (1971). Cells capable of colony formation in the peripheral blood of man. *Science*, **171**, 293–4.

14 Bagby, G. C., Lawrence, H. J., and Neerhout, R. C. (1983). T-lymphocyte-mediated granulopoietic failure. *In vitro* identification of prednisone-responsive patients. *N. Eng. J. Med.*, **309**, 1073–8.

15 Cudkowicz, G. and Stimpfling, J. H. (1964). Deficient growth of C57B1 mouse marrow cells transplanted in F1 hybrid mice. Association with the histocompatibility-2 locus. *Immunology*, **7**, 291–306.

16 Cudkowicz, G. and Bennett, M. (1971). Peculiar immunobiology of bone marrow allografts. I. Graft rejection by heavily "responder" mice. *J. Exp. Med.*, **134**, 83–102.

17 Cudkowicz, G. and Hochman, P. S. (1979). Do natural killer cells engage in regulated reaction against self to ensure homeostasis? *Immunol. Rev.*, **44**, 13–41.

18 Kiessling, R., Hochman, P. S., Haller, O., Shearer, G. M., Wigzell, H., and Cudkowicz, G. (1977). Evidence for a similar or common mechanism for natural killer cell activity and resistance to hematopoietic grafts. *Eur. J. Immunol.*, **7**, 655–63.

19 Daley, J. P., Wroblewski, J. M., Kaminsky, S. G., and Nakamura, I. (1987). Genetic control of the target structures recognized in hybrid resistance. *Immunogenetics*, **26**, 21–30.

20 Rembecki, R. M., Kumar, V., David, C. S., and Bennett, M. (1988). Bone marrow cell transplants involving intra-H-2 recombinant inbred mouse strains. Evidence that hematopoietic histocompatibility-1 (Hh-1) genes are distinct from H-2D or H-2L. *J. Immunol.*, **141**, 2253–60.

21 Öhlén, C., *et al.* (1989). Prevention of allogeneic bone marrow graft rejection by H-2 transgene in donor mice. *Science, 246,* 666–8.

22 Okumura, K., Habu, S., and Shimamura, K. (1982). The role of asialo GM1+ (GA1+) cells in the resistance to transplants of bone marrow or other tissues. In *NK cells and other natural effector cells* (ed. R. B. Herberman), pp. 1527–33. Academic Press, New York.

23 Lotzova, E., Pollack, S. B., and Savary, C. A. (1982). Direct evidence

for the involvement of natural killer cells in bone marrow transplantation. In *NK cells and other natural effector cells* (ed. R. B. Herberman), pp. 1535–40. Academic Press, New York.

24 Warner, J. F. and Dennert, G. (1982). Effects of a cloned cell line with NK activity on bone marrow transplants, tumour development and metastasis *in vivo*. *Nature*, **300**, 31–4.

25 Yankelevich, B., Knobloch, C., Nowicki, M., and Dennert, G. (1989). A novel cell type responsible for marrow graft rejection in mice. T cells with NK phenotype cause acute rejection of marrow grafts. *J. Immunol.*, **142**, 3423–30.

26 Murphy, W. J., Kumar, V., and Bennett, M. (1987). Rejection of bone marrow allografts by mice with severe combined immunodeficiency (SCID). Evidence that natural killer cells can mediate the specificity of marrow graft rejection. *J. Exp. Med.*, **165**, 1212–17.

27 Warner, J. F. and Dennert, G. (1985). Bone marrow graft rejection as a function of antibody-directed natural killer cells. *J. Exp. Med.*, **161**, 563–76.

28 Bordignon, C., Daley, J. P., and Nakamura, I. (1985). Hematopoietic histoincompatibility reactions by NK cells *in vitro*: model for genetic resistance to marrow grafts. *Science*, **230**, 1398–1401.

29 Ciccone, E., *et al.* (1988). Specific lysis of allogeneic cells after activation of CD3⁻ lymphocytes in mixed lymphocyte culture. *J. Exp. Med.*, **168**, 2403–8.

30 Ciccone, E., *et al.*, (1990). Specific recognition of human CD3⁻ CD16⁺ natural killer cells requires the expression of an autosomic recessive gene on target cells. *J. Exp. Med.*, **172**, 47–52.

31 Ferrara, J. L., Mauch, P., Van Dijken, P. J., Crosier, K. E., Michaelson, J., and Burakoff, S. J. (1990). Evidence that anti-asialo GM1 *in vivo* improves engraftment of T cell-depleted bone marrow in hybrid recipients. *Transplantation*, **49**, 134–8.

32 Murphy, W. J., Kumar, V., Cope, J. C., and Bennett, M. (1990). An absence of T cells in murine bone marrow allografts leads to an increased susceptibility to rejection by natural killer cells and T cells. *J. Immunol.*, **144**, 3305–11.

33 Welsh, R. M. (1978). Cytotoxic cells induced during lymphocytic choriomeningitis virus infection of mice. I. Characterization of natural killer cell induction. *J. Exp. Med.*, **148**, 163–81.

34 Welsh, R. M. (1986). Regulation of virus infections by natural killer cells. A review. *Nat. Immun. Cell Growth Regul.*, **5**, 169–99.

35 Biron, C. A. and Welsh, R. M. (1982). Blastogenesis of natural killer cells during viral infection *in vivo*. *J. Immunol.*, **129**, 2788–95.

36 Bro-Jorgensen, K. (1978). The interplay between lymphocytic choriomeningitis virus, immune function, and hemopoiesis in mice. *Adv. Virus Res.*, **22**, 327–69.

37 Bro-Jorgensen, K. and Knudtzon, S. (1977). Changes in hemopoiesis during the course of the acute LCM virus infection in mice. *Blood, 49,* 47–57.

38 Randrup-Thomsen, A., Pisa, P., Bro-Jorgensen, K., and Kiessling, R. (1986). Mechanisms of lymphocytic choriomeningitis virus-induced hemopoietic dysfunction. *J. Virol., 59,* 428–33.

39 Young, N. S., Mortimer, P. P., Moore, J. G., and Humphries, R. K. (1984). Characterization of a virus that causes transient aplastic crisis. *J. Clin. Invest., 73,* 224–30.

40 Young, N. and Mortimer, P. (1984). Viruses and bone marrow failure. *Blood., 63,* 729–37.

41 Qiu, F., *et al.* (1988). Primary structure of c-kit: relationship with the CSF-1/PDGF receptor kinase family. Oncogenetic activation of c-kit involves deletion of extracellular domain and C-terminus. *Embo J., 7,* 1003–11.

42 Fleishman, R. A., Custer, R. P., and Mintz, B. (1982). Totipotent haematopoietic stem cells: normal self-renewal and differentiation after transplantation between mouse fetuses. *Cell., 30,* 351–9.

43 Landreth, K. S., Kincade, P. W., Lee, G., and Harrison, D. E. (1984). B lymphocyte precursors in embryonic and adult W anemic mice. *J. Immunol., 132,* 2724–9.

44 Hansson, M., Petersson, M., Koo, G. C., Wigzell, H., and Kiessling, R. (1988). *In vivo* function of natural killer cells as regulators of myeloid precursor cells in the spleen. *Eur. J. Immunol., 18,* 485–8.

45 Pantel, K. and Nakeff, A. (1989). Differential effect of natural killer cells on modulating CFU-Meg and BFU-E proliferation *in situ. Exp. Hematol., 17,* 1017–21.

46 Pantel, K., Boertman, J., and Nakeff, A. (1990). Inhibition of haematopoietic recovery from radiation-induced myelosuppression by natural killer cells. *Radiat Res., 122,* 168–71.

47 Bordignon, C., *et al.* (1989). Graft failure after T-cell-depleted human leukocyte antigen identical marrow transplants for leukemia: II. *In vitro* analyses of host effector mechanisms. *Blood, 74,* 2237–43.

48 Vinci, G., *et al.* (1988). *In vitro* inhibition of normal human hematopoiesis by marrow CD3$^+$, CD8$^+$, HLA-Dr$^+$, HNK1$^+$ lymphocytes. *Blood, 72,* 1616–21.

49 Morikawa, K., Nakano, A., Oseko, F., and Morikawa, S. (1989). HIgh cytotoxic cell activity in the marrow from patients with aplastic anemia. *Jap. J. Med., 28,* 585–92.

50 Abdou, N. I., Verdirame, J. D., Amare, M., and Abdou, N. L. (1981). Heterogeneity of pathogenetic mechanisms in aplastic anemia. *Ann. Int. Med., 95,* 43–50.

51 Doney, K. C., Weiden, P. L., Buckner, C. D., Storb, R., and Thomas, E. D. (1981). Treatment of severe aplastic anemia using antithymocyte

globulin with or without an infusion of HLA haploidentical marrow. *Exp. Hematol.*, **9**, 829–33.

52 Rothmann-Hamburger, S. A., Bukowski, R. M., Finke, J. A., Laffay, D. L., Hoffman, G. L., and Henlett, J. S. (1979). Evidence for immune mediated aplastic anemia and its treatment with anti-thymocyte globulin. *Clin. Res.*, **27**, 296–302.

53 Goss, G. D., *et al.* (1985). Effect of natural killer cells on syngeneic bone marrow: *in vitro* and *in vivo* studies demonstrating graft failure due to NK cells in an identical twin treated by bone marrow transplantation. *Blood.*, **66**, 1043–6.

54 Freimark, B., Lanier, L., Phillips, J., Quertermous, T., and Fox, R. (1987). Comparison of T cell receptor gene rearrangements in patients with large granular T cell leukemia and Felty's syndrome. *J. Immunol.*, **138**, 1724–9.

55 Loughran, T. P. J., Clark, E. A., Price, T. H., and Hammond, W. P. (1986). Adult-onset cyclic neutropenia is associated with increased large granular lymphocytes. *Blood.*, **68**, 1082–87.

56 Iizuka, Y., Nishinarita, S., Ohshima, T., and Sawada, S. (1989). Hematopoiesis in patients with acute natural killer cell leukemia and large granular lymphocytosis: relationship between clinical features and haematopoietic inhibitor activity of peripheral mononuclear cells [*letter*]. *Eur. J. Haematol.*, **43**, 257–8.

57 Grillot-Courvalin, C., Vinci, G., Tsapis, A., Dokhelar, M. C., Vainchenker, W., and Brouet, J. C. (1987). The syndrome of T8 hyperlymphocytosis: variation in phenotype and cytotoxic activities of granular cells and evaluation of their role in associated neutropenia. *Blood*, **69**, 1204–10.

58 Kiessling, R. and Wigzell, H. (1981). Surveillance of primitive cells by natural killer cells. *Curr. Top. Microbiol. Immunol.*, **92**, 107–23.

59 Hansson, M., Kiessling, R., and Andersson, B. (1981). Human fetal thymus and bone marrow contain target cells for natural killer cells. *Eur. J. Immunol.*, **11**, 8–12.

60 Hansson, M., Kiessling, R., Andersson, B., Karre, K., and Roder, J. (1979). Natural killer (NK) sensitive T-cell subpopulation in the thymus: inverse correlation to NK activity of the host. *Nature*, **278**, 174–6.

61 Riccardi, C., Santoni, A., Barlozzari, T., and Herberman, R. B. (1981). *In vivo* reactivity of mouse natural killer (NK) cells against normal bone marrow cells. *Cell Immunol.*, **60**, 136–43.

62 Barr, R. D. and Stevens, C. A. (1982). The role of autologous helper and suppressor T cells in the regulation of human granulopoiesis. *Am. J. Hemat.*, **12**, 323–6.

63 Morris, T. C. M., Vincent, P. C., Sutherland, R., and Hersey, P. (1980). Inhibition of normal granulopoiesis *in vitro* by non-B non-T lymphocytes. *Br. J. Haematol.*, **45**, 541–50.

64 Hansson, M., Beran, M., Andersson, B., and Kiessling, R. (1982). Inhibition of *in vitro* granulopoiesis by autologous and allogeneic human NK cells. *J. Immunol.*, **129**, 126–32.

65 Spitzer, G. and Verma, D. S. (1982). Cells with Fc receptors from normal donors suppress granulocyte-macrophage colony formation. *Blood*, **60**, 758–66.

66 Matera, L., Santoli, D., Garbarino, G., Pegoraro, L., Bellone, G., and Pagliardi, G. (1986). Modulation of *in vitro* myelopoiesis by LGL: different effects on early and late progenitor cells. *J. Immunol.*, **136**, 1260–5.

67 Mangan, K. F., (Chikkappa, G., Bieler, L. F., Scharfman, W. B., and Parkinson, D. R. (1982). Regulation of human blood erythroid burst-forming unit (BFU-E) proliferation by T-lymphocyte subpopulations defined by Fc receptors and monoclonal antibodies. *Blood*, **59**, 990–6.

68 Degliantoni, G., Perussia, B., Mangoni, L., and Trinchieri, G. (1985). Inhibition of bone marrow colony formation by human natural killer cells and by natural killer cell-derived colony-inhibiting activity. *J. Exp. Med.*, **161**, 1152–68.

69 Degliantoni, G., Murphy, M., Kobayashi, M., Francis, M-K., Perussia, B., and Trinchieri, G. (1985). Natural Killer (NK) cell-derived hematopoietic colony-inhibiting activity and NK cytotoxic factor. Relationship with tumor necrosis factor and synergism with immune interferon. *J. Exp. Med.*, **162**, 1512–30.

70 Herrmann, F., Schmidt, R. E., Ritz, J., and Griffin, J. D. (1987). *In vitro* regulation of human hematopoiesis by natural killer cells: analysis at a clonal level. *Blood*, **69**, 246–54.

71 Lipton, J. M., Kudisch, M., Gross, R., and Nathan, D. (1986). Defective erythroid progenitor differentiation system in congenital hypoplastic (Diamond–Blackfan) anemia. *Blood*, **67**, 962–8.

72 Nagler, A. and Greenberg, P. L. (1990). Bone marrow cell modulation and inhibition of myelopoiesis by large granular lymphocytes and natural killer cells. *Int. J. Cell Cloning*, **8**, 171–83.

73 Nagler, A., Greenberg, P. L., Lanier, L. L., and Phillips, J. H. (1988). The effects of recombinant interleukin 2-activated natural killer cells on autologous peripheral blood hematopoietic progenitors. *J. Exp. Med.*, **168**, 47–54.

74 Niemeyer, C. M., Sieff, C. A., Smith, B. R., Ault, K. A., and Nathan, D. G. (1989). Hematopoiesis *in vitro* coexists with natural killer lymphocytes. *Blood*, **74**, 2376–82.

75 Van Den Brink, M. R., Voogt, P. J., Marijt, W. A., Van Luxemburg Heys, S. A., Van Rood, J. J., and Brand, A. (1989). Lymphokine-activated killer cells selectively kill tumor cells in bone marrow without compromising bone marrow stem cell function *in vitro*. *Blood*, **74**, 354–60.

76 Voogt, P. J., *et al.* (1989). Normal haematopoietic progenitor cells and malignant lymphohematopoietic cells show different susceptibility to direct cell-mediated MHC-restricted lysis by T cell receptor$^-$/CD3$^-$, T cell receptor $\gamma\delta^+$/CD3$^+$ and T cell receptor $\alpha\beta^+$/CD3$^+$ lymphocytes. *J. Immunol.*, **142**, 1774–80.

77 Perussia, B., Trinchieri, G., and Cerottini, J. C. (1979). Functional studies of Fc receptor-bearing human lymphocytes: effect of treatment with proteolytic enzymes. *J. Immunol.*, **123**, 681–7.

78 Anegón, I., Cuturi, M. C., Trinchieri, G., and Perussia, B. (1988). Interaction of Fcγ receptor (CD16) with ligands induces transcription of IL-2 receptor (CD25) and lymphokine genes and expression of their products in human natural killer cells. *J. Exp. Med.*, **167**, 452–72.

79 Cuturi, M. C., *et al.* (1989). Production of haematopoietic colony-stimulating factors by human natural killer cells. *J. Exp. Med.*, **169**, 569–83.

80 Gewirtz, A. M., Xu, W. Y., and Mangan, K. F. (1987). Role of natural killer cells, in comparison with T lymphocytes and monocytes, in the regulation of normal human megakaryocytopoiesis *in vitro*. *J. Immunol.*, **139**, 2915–24.

81 Scala, G., *et al.* (1984). Human large granular lymphocytes are potent producers of interleukin-1. *Nature*, **309**, 56–9.

82 Trinchieri, G., Matsumoto-Kobayashi, M., Clark, S. C., Sheehra, J., London, L., and Perussia, B. (1984). Response of resting human peripheral blood natural killer cells to interleukin-2. *J. Exp. Med.*, **160**, 1147–69.

83 Ortaldo, J. R., *et al.* (1984). Effects of natural and recombinant IL 2 on regulation of IFNγ production and natural killer activity: lack of involvement of the Tac antigen for these immunoregulatory effects. *J. Immunol.*, **133**, 779–83.

84 Pistoia, V., Cozzolino, F., Torcia, M., Castigli, E., and Ferrarini, M. (1985). Production of B cell growth factor by a Leu7$^+$, OKM1$^+$ non-T cell with the features of large granular lymphocytes (LGL). *J. Immunol.*, **134**, 3179–84.

85 Linch, D. C., Lipton, J. M., and Nathan, D. G. (1985). Identification of three accessory cell populations in human bone marrow with erythroid burst-promoting properties. *J. Clin. Invest.*, **75**, 1278–84.

86 Pistoia, V., Ghio, R., Nocera, A., Leprini, A., Perata, A., and Ferrarini, M. (1985). Large granular lymphocytes have a promoting activity on human peripheral blood erythroid burst-forming units. *Blood*, **65**, 464–72.

87 Witmer-Pack, M. D., Olivier, W., Valinsky, J., Schuler, G., and Steinman, R. M. (1987). Granulocyte/macrophage colony-stimulating factor is essential for the viability and function of cultured murine epidermal Langerhans cells. *J. Exp. Med.*, **166**, 1484–98.

88 Kasahara, T., Djeu, J. Y., Dougherty, S. F., and Oppenheim, J. S. (1983). Capacity of human large granular lymphocytes (LGL) to produce multiple lymphokines: interleukin 2, interferon and colony stimulating factor. *J. Immunol.,* **131,** 2379–85.

89 Pistoia, V., *et al.* (1989). Production of colony-stimulating activity by human natural killer cells: analysis of the conditions that influence the release and detection of colony-stimulating activity. *Blood,* **74,** 156–64.

90 Murphy, M., Perussia, B., and Trinchieri, G. (1988). Effects of recombinant tumor necrosis factor, lymphotoxin and immune interferon on proliferation and differentiation of enriched haematopoietic precursor cells. *Exp. Hematol.,* **16,** 131–8.

91 Murphy, M., Loudon, R., Kobayashi, M., and Trinchieri, G. (1986). Gamma interferon and lymphotoxin, released by activated T cells, synergize to inhibit granulocyte-monocyte colony formation. *J. Exp. Med.,* **164,** 263–79.

92 Trinchieri, G. Effects of tumor necrosis factor and lymphotoxin on the hematopoietic system. In *Tumor necrosis factors: Structure, function and mechanisms of action* (ed. B. B. Aggarwal and J. Vilcek) Marcel Dekker, New York. (In press.)

93 Broxmeyer, H. E., *et al.* (1986). The suppressive influences of human tumor necrosis factors on bone marrow hematopoietic progenitor cells from normal donors and patients with leukemia: synergism of tumor necrosis factor and interferon-γ. *J. Immunol.,* **136,** 4487–95.

94 Caux, C., Saeland, S., Favre, C., Duvert, V., Mannoni, P., and Banchereau, J. (1990). TNF alpha strongly potentiates IL-3 and GM-CSF induced proliferation of human CD34$^+$ hematopoietic progenitor cells. *Blood,* **75,** 2292–8.

95 Trinchieri, G., Kobayashi, M., Rosen, M., Loudon, R., Murphy, M., and Perussia, B. (1986). Tumor necrosis factor and lymphotoxin induce differentiation of human myeloid cell lines in synergy with immune interferon. *J. Exp. Med.,* **164,** 1206–25.

96 Means, R. T., Jr., Dessypris, E. N., and Krantz, S. B. (1990). Inhibition of human colony-forming-unit erythroid by tumor necrosis factor requires accessory cells. *J. Clin. Invest.,* **86,** 538–41.

97 Neta, R. and Oppenheim, J. J. (1988). Cytokines in therapy of radiation injury. *Blood,* **72,** 1093–95.

98 Slordal, L., Warren, D. J., and Moore, M. A. S. (1989). Effect of recombinant murine tumor necrosis factor on haematopoietic reconstitution in sublethally irradiated mice. *J. Immunol.,* **142,** 833–5.

99 Kohase, M., Henriksen-Destefano, D., May, L. T., Vilcek, J., and Sehgal, P. B. (1986). Induction of beta$_2$ interferon by tumor necrosis factor: a homeostatic mechanism in the control of cell proliferation. *Cell,* **45,** 659–66.

100 Seelentag, W., Mermod, J.-J., and Vassalli, P. (1989). Interleukin 1 and tumor necrosis factor-alpha additively increase the levels of granulocyte-macrophage and granulocyte colony-stimulating factor (CSF) in mRNA in human fibroblasts. *Eur. J. Immunol.*, **19**, 209–12.

101 Munker, R., Gasson, J., Ogawa, M., and Koeffler, H. P. (1986). Recombinant human TNF induces production of granulocyte-monocyte colony-stimulating factor. *Nature*, **323**, 79–82.

102 Cassatella, M. A., Anegón, I., Cuturi, M. C., Griskey, P., Trinchieri, G., and Perussia, B. (1989). FcγR(CD16) interaction with ligand induces Ca^{2+} mobilization and phosphoinositide turnover in human natural killer cells. Role of Ca^{2+} in FcγR(CD16)-transcription and expression of lymphokine genes. *J. Exp. Med.*, **169**, 549–67.

103 Beran, M., Hansson, M., and Kiessling, R. (1983). Human natural killer cells can inhibit clonogenic growth of fresh leukemic cells. *Blood*, **61**, 596–99.

104 Long, G. S., Hiserodt, J. C., Harnaha, J. B., and Cramer, D. V. (1988). Lymphokine activated killer cell purging of leukemia cells from bone marrow prior to syngeneic transplantations. *Transplantation*, **46**, 433–8.

3 Natural killer cells in bacterial infection

P. GARCIA-PEÑARRUBIA

1   Introduction

Natural killer (NK) cells are morphologically defined as large granular lymphocytes (LGLs) of a variable phenotype, which usually includes the expression of the antigens, CD56 (Leu19; NKH1) and CD16 (Leu11; Fc for IgGIII), and the invariant zeta chain component of the T-cell receptor (1–6). NK cells constitute up to 15 per cent of peripheral blood lymphocytes of healthy donors, although there are large individual variations (3). As discussed more fully in the first chapter of this volume, NK cells can also be defined functionally as cells that mediate non-histocompatibility restricted killing of certain targets, in which effector cells are derived from hosts not specifically immunized by target antigen, hence the term 'natural killing' (1, 3, 7–9). These cells have been implicated in immunosurveillance mechanisms against malignancy because of their ability to promiscuously kill a variety of tumour targets (7, 8), particularly when activated by cytokines (10–14), although their *in vivo* role in malignancy remains unclear. The activity of NK cells against infectious diseases has received less attention, although in recent years increasing evidence has emerged that NK cells are involved in defence against such forms of disease (15–20). Thus, cells with NK activity have been able to inhibit microbial colonization and growth, including intracellular and extracellular parasites, fungi, and a wide variety of viral infections (15–20). In contrast, the role played by NK cells in host defences against bacterial infections has remained unclear, and sometimes controversial (20–21). The aim of this article is to review the evidence for, and the significance of, the antibacterial activity of NK cells, as well as the effect that bacteria and bacterial infections exert on NK cell activity.

2   Indirect evidence for the antibacterial activity of NK cells against various forms of bacteria

2.1   Gram-negative bacteria

The first suggestion of natural lymphoid antibacterial activity was made by Lowell and co-workers (22) who showed that human peripheral lymphoid cells are able to kill *Neisseria meningitidis* (22) and *Shigella flexneri* (23) in the presence of immune sera, presumably by an antibody-dependent cellular cytotoxicity (ADCC) mechanism. Later, Nencioni and co-workers (24) found low levels of natural cytotoxic activity against *Salmonella typhimurium* among fresh, non-immune lymphocytes from murine spleen and intestinal lamina propia. These cells were asialo-GM$_1^+$, FcR$^+$, and Thy1.2$^-$, a phenotype which is compatible with that of NK cells. These authors also showed that secretory IgA could mediate antibacterial ADCC when bound to lymphocytes from immune gut-associated lymphoid tissues (GALT) (25). However, by studying natural antibacterial activity of human peripheral blood lymphocytes (PBL) against *Salmonella typhi* this property was ascribed to T4$^+$ cells armed with IgA antibodies (26). This activity was low even at high effector to target (E/T) ratios (Table 3.1), but was increased after oral immunization with the live *S. typhi* mutant strain Ty21a, thereby indicating that this kind of antibacterial activity is not completely non-specific (27). Recently, these authors also studied the natural antibacterial activity of peripheral blood mononuclear cells (PBMC) against *S. typhi* in HIV-infected patients (28). As expected, they found low percentages of CD4$^+$ lymphocytes, the value of the CD4/CD8 ratio always being lower than 1, and the natural antibacterial activity against *S. typhi* and *paratyphi C* was completely absent. When the pheno-type of the cells exerting this antibacterial activity was characterized by depleting different effector cell populations with the specific monoclonal antibody plus complement, it was found that natural antibacterial activity was abolished after depletion of CD2$^+$, CD3$^+$, and CD4$^+$ cells, but it was increased after depletion of CD8$^+$, CD11$^+$, and CD16$^+$ cells. Thus, these authors concluded that the effector cell responsible for natural antibacterial activity was a non-monocytic, non-NK CD2$^+$, CD3$^+$, or CD4$^+$ T-cell. In contrast, the antibacterial activity of newborn cord blood was greatly de-creased by depletion of CD3$^+$ or CD11$^+$ cells, suggesting that natural antibacterial activity in the newborn is expressed by monocytes and a subset of CD4$^+$ T-cells (29). This activity was also exerted through an ADCC mechanism in which the effector cells were armed with pre-existing maternal IgG antibodies instead of IgA. However, one representative experiment in the same study [see (29)] shows that the antibacterial activity was also inhibited when the effector population was depleted of CD16$^+$ cells (although to a lesser extent than if the CD4$^+$ population was suppressed) and therefore the activity decreased 74.4 per cent at 50/1 ratio, 33.72 per cent at 100/1 ratio, and 6.52 per cent at 200/1 ratio. Hence, this study did not rule out the possibility that NK cells also exerted antibacterial activity.

Glutaraldehyde-fixed *Salmonella* bacteria and bacterial cell walls are able to activate NK cells to enhanced tumour target killing (30). PBL depleted of macrophages by adherence to glass and nylon wool were occasionally bactericidal at high effector to target ratios (e.g. 100/1) against the smooth and rough *S. typhimurium* strains used to activate NK cells (Table 3.1). However, this activity was ascribed to macrophages contaminating the PBL population and so it was concluded that activation of cytotoxic activity by bacterial products was not parallel to the expression of antibacterial activity. All inducible cytotoxic activity was in the subset of lymphocytes expressing Leu19 (NKH1) antigen (31). In addition, this

Table 3.1 Antibacterial effect of human natural killer cells

Donor	Cells	Antibacterial activity (%)									Bacteria	Ref.
		0.2^a	1	5	10	15	25	50	100	200		
Healthy	MDC							18	30	37	*S.typhi*	26
Healthy[b]	MDC					24	43	50			*S.typhi*	26
Healthy	PBMC						−17	5	23	30	*S.typhi*	28
LAS/ARC	PBMC						−13	−21	−25	−10	*S.typhi*	28
AIDS	PBMC						−21	−9	2		*S.typhi*	28
Cord blood	PBMC						25	33	60	72	*S.typhi*	29
Healthy	Percol F_2								4;19		*S.typhimurium* (Ra)	30
Healthy	PBMC								32		*S.typhimurium* (Ra)	30
Healthy	Percoll F_2								3;19		*S.typhimurium* (S-4,12)	30
Healthy	PBMC								19		*S.typhimurium* (S-4,12)	30
Healthy	Percoll F_{2+3}	19									*S.aureus*	32
Healthy	MDC					15	23			41	*S.typhi*	109
Healthy	Percol F_2			11	27	45		53			*S.typhi*	109
Healthy	CD16$^+$		>85				>99.9	>99.9	>99.9	>99.9	*S.typhi*	109

LAS/ARC, lymphadenopathy syndrome/AIDS-related complex; MDC, monocyte-depleted cells; PBMC, peripheral blood mononuclear cells, Percoll F, fraction of Percoll gradient used to purify cells.
[a] Effector to target ratio.
[b] After oral live Ty21a *S.typhi* vaccine.

stimulation of cytotoxic potential was not dependent on their release of interferon (IFN) as there was no detectable IFN in the supernatants of NK cell cultures co-incubated in the presence of bacteria.

2.2   Gram-positive bacteria

In contrast to the above results, Abo and co-workers found that fresh NK cells purified on a discontinuous Percoll density gradient, or by fluorescence-activated cell sorting with the monoclonal marker, HNK1, were able to phagocytose Gram-positive (Gram+) bacteria and these cells also expressed a substantial level of bactericidal activity against *Staphylococcus aureus* [the mean value of inhibition was 19 per cent, Table 3.1 (32)]. Gram-negative (Gram−) bacteria and yeasts were not phagocytosed, and no antibacterial assays against these micro-organisms were reported. The stimulation of LGLs by phagocytosis has been shown to induce their production of IL-1 and IFN, but not IL-2. Whereas non-phagocytic particles, such as latex or silica, did not promote cytokine synthesis. However, these authors also described that Gram− bacteria, such as *Escherichia coli* or *Salmonella minnesota*, can induce the production of IL-1-like factors to a level comparable to that of Gram+ bacteria, although NK cells were not able to engulf these micro-organisms. The results of this study show that some components of bacterial cell walls are important as recognition structures for NK cells since treatment of bacteria with lysostaphin reduced both the percentage of cells phagocytosing *S. aureus* and the production of IL-1-like factors. Abo and co-workers postulated that NK cells could belong to two different lines, one acting as a T-cell accessory by producing IL-2, and the other having myelocytic characteristics, that is, bacterial recognition, phagocytosis, production of IL-1, and, probably as suggested by Greenberg (21), acting as an antigen-presenting cell. Tagliabue's group has also described natural antibacterial activity from murine lung lymphocytes against *Streptococcus pneumoniae* (33), but this activity was also ascribed to helper T-cells through an IgA-driven mechanism (34). However, in contrast to what could be expected in the assumption that NK cells play a role in defence against bacteraemia, a recent survey of patients with advanced cancer treated by known NK cell stimulants, including IL-2, IFNα, and TNFα, did not report any protection against bacterial infections (35). The incidence of infection in these patients was 23 per cent, with the most frequent micro-organisms isolated being such Gram+ bacteria as *Staphylococcus* and *Streptococcus*. However, the study was only performed on a small group of patients and a control population without treatment was not evaluated. Hence, further studies are needed to confirm these results before any conclusion can be drawn.

2.3   Intracellular bacteria

The role played by NK cells in defence against intracellular bacterial
infections is also unclear. Klimpel and co-workers (36) showed that lymphocytes with NK cell characteristics were able to kill *S. flexneri*-infected
HeLa cells (36). This activity was dependent upon bacterial invasion of the
HeLa cells and was greatly enhanced by pretreatment of PBL with IL-2 or
IFNα (37). They also showed that co-culture of PBL with *Salmonella* or
Shigella greatly increased their cytotoxic activity against K562 tumour cells
and *S. flexneri*-infected HeLa cells (38). That bacteria-enhanced NK cell
activity required the direct contact of bacteria (either heat killed or viable)
with cells was demonstrated by the fact that lipopolysaccharide (LPS) and
supernatants from PBL-bacteria cultures lacking IFN activity did not enhance NK cell activity. Moreover, incubation of PBL with *Salmonella* or
Shigella for 18 h promoted high levels of antiviral activity which appeared
to be mediated by a mixture of IFNα and IFNγ. Although these authors
did not show evidence of direct extracellular antibacterial activity, they
postulated that bacteria-induced synthesis of IFN could be an important
mechanism of defence against bacterial infections, since it was previously
found that pretreatment of HeLa cells or mouse fibroblasts with IFNα or
IFNγ results in a refractory state to subsequent bacterial invasion by
Salmonella or *Shigella* (37). Furthermore, other authors have demonstrated that IFN is able to inhibit intracellular pathogens, such as *Rickettsia*
(39), *Chlamydia* (40), *Legionella* (41), and *Toxoplasma gondii* (42). A
recent report has also reported that PBL and non-adherent cells from non-
immune donors can lyse autologous *Legionella*-infected monocytes, and
that cytolytic activity was increased by IL-2 activation of effector cells. The
phenotype of the effector cells belonged to the OKM1[+], OKT11[+], and
Leu11[+] population of LGLs (43). These authors suggested that the elimination of the infected monocytes may limit the course of human Legionel-
losis since lysis of the host cell interfered with bacterial replication. Later,
they also found that *Legionella* could induce the production of IFNγ, both
in vitro and *in vivo* (44), by spleen asialo-GM$_1^+$ or NK-like cells (44). In
addition, *Legionella pneumophila* antigens were able to stimulate NK
activity against YAC1 target cells, both in PBL and lung leucocytes, after
an 18 h period of co-culture of splenocytes *in vitro*, or during *in vivo*
infection of intratracheal inoculated mice. There also seemed to be a
mobilization and recruitment of NK cells in the lungs of infected mice.
Thus, the total number of leucocytes recovered from infected lungs was
20- to 25-fold higher than that recovered from uninfected mice. Similar
results for the ability of human NK cells to lyse *Mycobacterium avium*
complex-infected monocytes have recently been reported (45). Lysis of
infected cells paralleled the decrease in the viability of the micro-

organism. These authors also reported that the release of lymphocyte-derived mediators, such as TNFα, IL-2, and IFNα or IFNγ, could not be implicated as a cause of monocyte death. Contrary to these results, Zychlinsky and co-workers (46) have described the inability of a highly homogeneous population of murine lymphokine-activated killer (LAK) cells to kill macrophages infected with the intracellular agents, *M. avium, L. monocytogenes, L. pneumophila, T. gondii* or *Trypanosoma cruzi*, although they were efficient killers of the murine NK target, YAC1.

Another facultative intracellular bacterium, *Listeria monocytogenes,* has been reported to mobilize NK cells to the peritoneal cavity of mice early after injection of bacteria (47). Furthermore, this micro-organism can enhance NK cell activity *in vitro* and the cytotoxicity of NK cells isolated from infected animals (48– 49). Similarly, the mouse pneumonitis agent, *Chlamydia trachomatis*, is able to increase early NK activity in spleen and lung in pulmonary infections. The significance of these findings is not clear since depletion of NK activity with anti-asialo-GM$_1$ antibody, or NK stimulation by immunomodulators, did not affect the survival of animals or the counts of *C. trachomatis* (50). Also, biologically active TNFα was detected in the lung in nu/+ and in nu/nu mice after inoculation, and anti-TNFα antibody accelerated mortality and increased *C. trachomatis* counts in the lungs of infected animals (51). Thus, it was suggested that since TNFα is involved in a T-cell-independent pathway of macrophage activation (52), this pathway could be mediated by NK cells leading to the T-cell-independent production of IFNγ which is inhibitory for *C. trachomatis* (53). Also, it is known that TNFα is produced directly by NK cells (54), and that TNFα is able to exert not only antiviral activity (54), but also antibacterial resistance against *L. monocytogenes* (55, 56), and *Escherichia coli* in a susceptible mouse strain when combined with IL-1α (57), and against *Rickettsia conorii*, either alone or in combination with IFNγ (58). These studies raise the possibility that NK cells do not exert a direct antibacterial role, but could indirectly influence bacterial growth by activating and recruiting other effector cells, such as neutrophils and monocytes/macrophages, through the release of cytokines or other unknown NK cell factors. In this context, it is interesting to note that a soluble factor from LGLs has been shown to trigger a respiratory burst in monocytes against tumour targets (59), as well as their intracellular lysis of *S. aureus* (60). Similarly, LGLs are also able to release a factor that influences the growth of *Candida albicans* by activating neutrophils against the fungus (61).

3   Augmentation of NK activity by bacterial products

Recently, there has been increasing evidence that bacteria and bacterial products are able to potentiate killing by lymphocytes. In fact, several bacterial products are well-known immunomodulatory agents. Thus, natural killer cytotoxicity can be induced directly by mycobacteria, BCG (62), and such mycobacterial antigens as purified protein derivative (PPD) (63). Also, peptidoglycans isolated from bacterial cell walls and some synthetic analogues can exert an immunostimulatory influence similar to that elicited by whole bacteria (64–66). This stimulatory effect has been ascribed to various effector cell populations, such as macrophages (67), T- and B-lymphocytes (68), and NK cells (66, 69–73). In the modulation of NK cell cytotoxicity by muramyl dipeptides there appears to be a strain-related responsiveness (69, 71, 73). Thus, some strains responded to this stimulatory effect, whereas C57B1 mice were resistant to stimulation. Also, enhancement of NK and ADCC responses by peptidoglycan monomer (PGM) was faster (but also more transient) in mice with constitutively weak NK activity (C57B1) than in strong reactors (CBA, C3H) (73).

Similarly, liposomes incorporating muramyltripeptide-phosphatidyl ethanolamine (MTP-PE), which is a lipophilic analogue of MDP (the minimally active subunit of *Mycobacterium tuberculosis* capable of replacing the adjuvant activity in Freund's complete adjuvant), increased NK cell cytotoxicity (74). However, this augmentation was only found using pulmonary and hepatic NK cells and was not observed for splenic or peritoneal NK cells (74). The mechanism by which NK activity is modulated by peptidoglycans and analogues is unknown, although it has been suggested that it may be exerted through production of activating cytokines (73–74).

Yokota and co-workers (75) have recently described the augmentative effect of the cell wall skeleton of *Nocardia rubra* (N-CWS) on LAK cell induction. The cytolytic activity was expressed in a non-adherent subpopulation of cells with positive phenotype for Thy1.2 and asialo-GM$_1$ antigens (that is, not typical CTL or NK cells). Combined therapy with N-CWS and IL-2 prolonged the survival of animals bearing LAK-sensitive Lewis lung carcinoma, but not the survival of animals bearing LAK-resistant tumours (75).

Staphylococcus aureus enterotoxins (SE), especially A and B, are potent mitogens for human and murine T-cells, as well as immunomodulators and inducers of IFNγ and other cytokines (76–81). The SEB stimulation of T-cells is strictly dependent on major histocompatibility complex (MHC) Class II-bearing cells (82–84). Recently, it has been described that SEB enhances natural cytotoxicity from human peripheral blood mononuclear cells. The phenotypes of precursor and effector cells of AKC (activated killer cytotoxicity) were CD16$^+$, CD5$^-$, and CD8$^-$, and the mechanism whereby SEB activates NK cells was IL-2-dependent since anti-IL2 antibody or cyclosporin inhibited the induction of AKC. Also, it was found that monocytes and IL-1 played no role in the generation of AKC (85). Indeed, we have also described that *in vitro* stimulation of monocyte-depleted peripheral blood mononuclear cells with SEB resulted in selective proliferation of cells with the phenotypic and functional characteristics of

NK cells (86). After culture for 4 to 5 days in the presence of SEB, 98–100 per cent of the cells expressed the CD16 antigen. This enrichment occurred through both the death of lymphocytes lacking NK cell markers and the proliferation of existing NK cells. These AKC cells can be maintained and expanded by IL-2 stimulation for 15 to 20 days. Activation of NK cells was detected by the appearance of receptors for IFNγ and IL-2, as well as by the augmentation of cytotoxicity against NK-susceptible and NK-resistant tumour cell lines and bacteria (86). Discrepancies between previous studies demonstrating the mitogenic effect of SEB on T-cells and the results obtained in our recent study (86), which indicated proliferation and activation of NK cells, could be due to the different culture conditions used, including the presence or absence of APC or MHC Class II-bearing accessory cells. The starting PBL population in this study (86) was depleted of monocytes and B-cells and, as discussed above, the mitogenic response of T-cells induced by SEB is strictly dependent on MHC Class II molecules (82–84).

There is an important body of evidence indicating the direct effects of LPS on various cell types; macrophages, B-cells, polymorphonuclear leucocytes, platelets, and certain T-cells (87–91). The effect of bacterial LPS on NK cell activity is controversial (30–31, 92–97). Some studies have found that LPS extracted from *Salmonella* bacteria interfered with or blocked lymphocyte activation mediated by whole Gram− bacteria (30–31). Conversely, Lindemann and co-workers have also described the stimulation of NK cell cytotoxicity by LPS isolated from periodontal bacteria (92). The level of bacterium-enhanced NK cytotoxicity at 24 h was comparable to that induced by IL-2, although by day 7 cytotoxicity of cultures with bacteria was 10-fold less than that of cultures exposed to the cytokine. The mechanism of NK cell activation by LPS is unknown. Salata and collaborators (95) have shown that LPS enhances NK activity by increasing both the conjugation of NK to target cells and the efficiency of killing mechanism(s). Other studies showed that bacterial stimulation required the synthesis of protein, but not DNA, and also that bacterial activation may occur by an IFN-independent mechanism. In turn, Lindemann (94) demonstrated that antibody to IFNα inhibited the activation of PBL cultured in the presence of LPS from enteric or periodontal bacteria, suggesting that IFNα may be, in part, responsible for NK cell activation. Moreover, high levels of IFN were measured in the supernatants of lymphocyte cultures stimulated with LPS. Anti-CAM antibodies (anticellular adhesion molecules) blocked cytotoxicity induced by LPS, but not IFNα, suggesting that anti-CAM antibodies have an indirect effect on NK activation by down-regulating LPS-induced IFNα production (94). Kang and others (96) have demonstrated both the adherence of LPS bilayers to the plasma membrane of human NK cells by 'edge attachment', and the LPS-enhancement of NK cell-mediated cytotoxicity. The latter author has also recently demonstrated the presence of Ca^{2+}-ATPase in the plasma membrane of human NK cells. The Ca^{2+} extruding activity of this enzyme was suppressed in cells exposed to LPS, but activated by the intracellular Ca^{2+} receptor, calmodulin. Thus, intracellular Ca^{2+} was increased in LPS-stimulated cells, but was decreased in cells treated with calmodulin (97). In addition, calmodulin was seen to block or reverse the inhibitory effect of LPS on the activity of this enzyme.

The kinetics of augmentation of NK activity in mice during the course of an active infection with *Pseudomonas aeruginosa* has been recently reported by Klein and Kearns (98). Following a sublethal injection of the bacterium, increased NK cell activity against YAC1 tumour cells was evident after 24 h, had peaked within 72 h, and returned to normal levels by 168 h. The route of administration directly influenced the augmentation of NK activity. Thus, the greatest increase was found in the lymphoid compartment most closely associated with the route of injection. Hence, these authors attributed the augmentation in NK activity to an influx of cells into the lymphoid compartment in which the *Pseudomona* was injected. The augmentation of NK activity was also elicited by a non-viable preparation of *P. aeruginosa*, suggesting that some components of the bacterium could attract or activate NK cytotoxicity. Furthermore, a recent study has shown that infection of mice with *Mycoplasma pulmonis* (MP) stimulated the cytotoxic activity of NK cells against YAC1 tumour cells *in vitro* and *in vivo* (99). NK cells of infected mice also directly inhibit colony formation of MP *in vitro* and eliminate viable MP from the lungs of i.v. infused animals. In addition, augmented NK cells in scid mice have been shown to inhibit MP growth in spleen and lung within 6 to 24 h after infusion; an effect that was abrogated by treatment with either anti-asialo-GM_1 serum or anti-IFNγ antibodies. Taken together, these reports support a positive role for NK cells in the host's natural resistance to bacterial infections.

4   Suppression of NK cell activity by bacterial products

It is known that infective micro-organisms produce substances able to suppress the host immune response in order to survive in an adverse environment (100–101). In fact, most of the bacteria that colonize the human oral mucosa can secrete substances that inhibit the *in vitro* proliferation of human peripheral blood mononuclear cells in response to T-cell mitogens (100).

Concerning the interaction between bacteria and NK cells, there are some reports of NK cell activity being inhibited by bacterial products (102–108). However, if the significance of NK cell recognition and activation by bacterial products is not clear, the relevance of NK cell inhibition by bacteria in the pathogenesis of bacterial diseases is totally unknown. In this context, NK cell suppression by immunomodulators like BCG (102), and *Corynebacterium parvum* has been described (103–106). The mechanism by which *C. parvum* inhibited NK cell activity involved erythroblasts competing with target cell binding, suppressor Thy1.2$^+$ lymphocytes (104–105), and also soluble substance(s) released by non-adherent cells from the peritoneal cavity of rats (106). The inhibitory effect was, therefore, not direct but exerted through induction of suppressor cells or suppressor factors released by other cell populations. In contrast, the extracellular products of *P. aeruginosa*, alkaline protease, and elastase, inhibited the cytotoxic activity of the NK cells against the tumour target cell, K562. That the proteases act directly on the effector cells was demonstrated when the inhibitory effect was seen not to be abrogated after removal of the proteases from effector cells by washing. Inhibition seems to occur via steric hindrance of specific NK cell receptors that recognize target structure (107). The same authors have also recently described a similar inhibitory effect on NK cell activity exerted through the cytotoxic protease of *L. pneumophila*. In this case, the inhibition of NK cell activity was observed at lower enzyme concentrations and after a shorter period of incubation than that induced by *P. aeruginosa*, alkaline protease, and elastase, suggesting a higher sensitivity of NK cells to this protease. *Legionella pneumophila* protease partly inhibited the binding of effector to target cells. The presence of the protease during the cytotoxicity assay was not required, indicating an effect of the protease at the receptor level (108). If as suggested (43–44), NK cells are involved in the clearance of *L. pneumophila*, the inhibition of NK cell activity by its protease could impair host defence against Legionnaire's disease.

Contrary to the results obtained by others (30–31, 38, 47–49, 93, 98), Abo and co-workers (32) have reported that co-incubation of NK cells with either phagocytosed (Gram+) or non-phagocytosed (Gram−) bacteria for 18 hours decreased their cytotoxic activity against K562 target cells.

5 Direct demonstration of *in vitro* antibacterial activity of NK cells

With these findings implicating NK cells in resistance to bacterial infections, we were interested to determine whether NK cells play a role in non-specific host defence mechanisms to infections with *Salmonella typhi* (109). To check this hypothesis NK cells were isolated from the blood of normal healthy volunteers using different procedures which yielded percentages of CD16$^+$ cells ranging from 15 to over 95 per cent (see Table 3.2). Thus, monocyte-depleted cells (MDC) were obtained using a Ficoll–Hypaque

Table 3.2 Surface phenotypes of lymphocyte populations

Group	Leu19[a]	Leu11	Leu1	LeuM3[b]
Leu11 cells prepared by negative selection[c]	99 ± 1	99 ± 1	0.2 ± 0	0
LGL Percoll cells	78 ± 1	79 ± 1	22 ± 2	0.1 ± 0.1
MDC	15 ± 1	15 ± 0.4	85 ± 1	0.1 ± 0.1
Leu11⁻ cells	0	0	72 ± 2	0

Phenotypes defined using specific FITC-labelled mononuclear antibodies (MABs).
[a] Leu19 antibody was conjugated with phycoerythrin (red) for FACS analysis.
[b] Leu19, Leu11, Leu1, and LeuM3 MABs were from Becton Dickinson & Co., and were used at 20 µl/ml in a 10^6 cell suspension.
[c] For obtaining the most purified population, MDCs were panned with the MABs Leu2, Leu3, and Leu4. [From Garcia-Peñarrubia *et al.* (109).]

gradient, adherence to glass petri dishes, and a nylon wool column. LGLs were prepared using discontinuous density Percoll gradient centrifugation as described previously (110). NK-enriched cells (Leu11$^+$) were prepared by the negative selection protocol illustrated in Fig. 3.1. The isolation procedure consists of panning the monocyte and B-cell-depleted population by incubation with unconjugated monoclonal antibodies (MABs), Leu2, Leu3, and Leu4, (20 µl/10^6 cells) for 30 min at 4 °C. They were then adhered to polystyrene plates coated with goat anti-mouse IgG for 60 min at 4 °C. The phenotype of the remaining, unattached NK-enriched cells was greater than 95 per cent Leu11$^+$ cells. In some experiments 100 per cent Leu19-enriched cells were obtained by fluorescence-activated cell sorting (FACS) and used. The bactericidal activity of these purified populations was assessed in assays against *Salmonella typhi*, *Escherichia coli*, and *Staphylococcus epidermidis* using standard procedures (109). Antibacterial activity was determined after 24 h incubation, comparing viable bacterial counts with control groups lacking NK cells. In parallel experiments, NK cytotoxic activity against the cell line K562 was studied using the standard 4 h ^{51}Cr release method (109).

5.1   Effect of NK cell enrichment

Results obtained in bactericidal assays with different lymphoid cell population against *S. typhi* showed that successive purification of Leu11 (CD16$^+$) cells resulted in significant increases in bactericidal activity (Fig. 3.2). Although antibacterial activity of MDC against *S. typhi* is dose-dependent, the maximum activity seldom rose above 40 per cent for the highest effector to target (E/T) ratio tested (200/1), which agrees with data from other sources (26, 28). The fractionation of lymphoid cells in Percoll gradients resulted in an important increase in the bactericidal activity

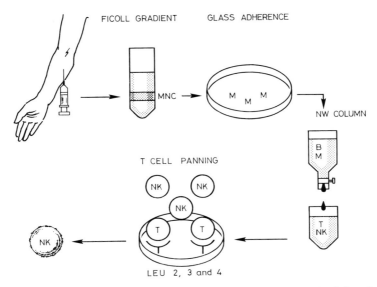

Fig. 3.1 Flow diagram of the isolation protocol used for obtaining Leu11-enriched NK cells. Peripheral blood from normal healthy donors was centrifuged on a Ficoll–Hypaque cushion to obtain mononuclear cells. Monocytes and B-cells were removed by successive adherence to glass petri dishes (twice) and a nylon wool column for 1 h at 37 °C. The resulting population was depleted of CD3$^+$, CD4$^+$, and CD8$^+$ T-cells by negative panning, involving incubation with monoclonal antibodies to Leu2, Leu3, and Leu4 (20 µg/10^6 cells) for 30 min at 4 °C, and adherence to polystyrene plates coated with goat anti-mouse IgG for 1 h at 4 °C.

compared to MDC. The inhibitory activity at 15/1 ratio was greater in all three experiments of Percoll-concentrated cells than in six experiments using MDC at the same ratio. In turn, Leu11-enriched cells obtained by panning showed the most potent antibacterial activity, (always greater than 80 per cent; that is, less than 20 per cent survival) at an E/T ratio of 10/1. There were no surviving bacteria at an E/T ratio of 20/1. These findings were confirmed using highly purified NK cells obtained by FACS (Fig. 3.2). In this case, bactericidal activity was greater than 99.9 per cent (that is, less than 0.1 per cent survival) at 1/1 ratio of cell/bacteria. In contrast the antibacterial activity of a cell population specifically depleted of Leu11$^+$ cells was similar to that in controls. These results constitute the first specific demonstration of the *in vitro* antibacterial activity of NK cells. The natural antibacterial activity of lymphoid cells was contained in the subpopulation bearing classical NK cell markers, CD16$^+$ and CD56$^+$. Thus, enrichment of lymphoid cells expressing those antigens was parallel to the expression of higher antibacterial activity. In addition, lymphoid cells depleted of CD16$^+$ cells were lacking in natural antibacterial activity.

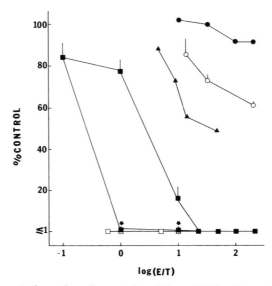

Fig. 3.2 Augmentation of antibacterial activity of NK cells against *S.typhi*, by successive enrichment of NK cells at different E/T ratios. Leu11-depleted cells (●), monocyte-depleted cells (○), Percoll-enriched (Percoll fractions 1 + 2 of MDC) (▲), Leu11/19 cells purified by panning (■), and Leu11/19 purified by cells sorting (□), were studied over a range of E/T values of 1/10–200/1. The most active Leu11-enriched population (∗) is shown separately from the other three experiments. Each point with standard deviation (s.d.) bars represents the mean of three separate experiments; other points represent one experiment. (From Garcia-Peñarrubia *et al.* (1989) (109).)

5.2   Effect of NK cell stimulation on antibacterial activity

Activation of NK cells was studied in two ways; simple incubation for 16 to 24 h at 37 °C, and pharmacological stimulation by staphylococcal enterotoxin B (SEB).

5.2.1   Effect of overnight incubation

Experiments depicted in Fig. 3.2 describe the behaviour of NK cells incubated for 16 h in HB104 media due to the duration of the cell purification procedures during the preceding day. To study the effect of 16 h incubation without addition of cytokine stimulation, MDC and Leu11-enriched populations were purified from a single donor, and their antibacterial activity was assessed either immediately or after 16 h incubation in HB104 media. Fresh MDC displayed insignificant antibacterial activity against all three bacterial targets (Fig. 3.3a), whereas Leu11-enriched cells showed substantial antibacterial activity, particularly against *E. coli*. After a 16 h incubation (Fig. 3.3b) the antibacterial activity of both cell populations was significantly enhanced against all bacteria targets tested. Although there was a modest increase in the number of Leu11$^+$ cells during the 16 h of incubation, this alone could not account for the increase in antibacterial activity. Rather, this could be attributed to an autocrine activation of Leu11-enriched cells, since fresh Leu11$^+$ cells from healthy donors did not express the activation markers, IL-2R and transferrin receptor, whereas after 16 h incubation 43-92 per cent express both receptors (A. D. Bankhurst, personal communication). Previous studies which found little or no antibacterial activity among murine (24–25) and human NK cells (28–31, 32; Table 3.1), used either unpurified lymphocyte populations (24–29) or freshly prepared monocyte-depleted Percoll-fractionated cells (30, 32). In order to characterize the phenotype of the effector cells in the antibacterial assays, depletion of cell subpopulations by MABs and complement has been used by other groups (28–29, 31). To this end, the change in antibacterial activity after depletion of just one specific lymphoid cell subpopulation was evaluated (26–29). This approach constitutes an indirect demonstration of the antibacterial activity exerted by that specific cell subpopulation. However, the global effect may be exerted by a mixed lymphoid population which can, in turn, interact with each other to activate or suppress their functional properties. In turn, positive selection procedures, such as the FACS, imply interaction of lymphoid CD receptors with their specific MABs, and this can itself result in alterations in cell functions. Conversely, our approach of negative selection to remove various cell types other than those expressing the NK cell phenotype, constitutes a direct demonstration of antibacterial activity exerted by human NK cells. Moreover, since we used in these studies highly purified NK cells, the problem of the interaction between different cell types was avoided. In any case, data shown in Table 3.1 demonstrate that our results are not in contradiction with those obtained by other groups. Thus, there are only small differences between the natural antibacterial activity of control PBMC found by Tagliabue and co-workers (26–28) and the corresponding values obtained by us using MDC populations. This could be due to small differences in the purity of the effector cells, and the stimulatory effect of an overnight incubation on NK activity (109). Also, the values of natural antibacterial activity obtained using PBMC from human cord blood are even higher than those found by us from MDC, although these authors ascribed this activity to CD4$^+$ T-cells armed with maternal IgG antibodies (29). The study which recorded 19 per cent inhibition of *S. aureus* by Percoll-concentrated cells (32) agrees with our results using the same cells and *S. typhi* targets. Tarkkanen and others (30) found very low levels of antibacterial activity in Percoll-fractionated cells at high E/T ratio (100/1), but attributed this activity to contaminating monocytes. However, our

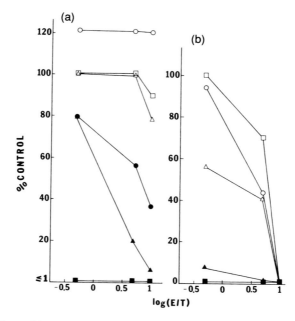

Fig. 3.3   Effect of *in vitro* overnight incubation in HB104 on antibacterial activity of MDC (*open symbols*) and Leu11-enriched cells (*filled symbols*) against *S.typhi* (○, ●), *E.coli* (UNM 101 strain) (□, ■), and *S.epidermidis* (△, ▲) at three different E/T ratios. (a) The antibacterial activity of fresh cells tested for 4–5 h after the blood was drawn. (b) The results obtained with the same cell population 24 h later. The cells were washed in HB104 before mixing with bacteria. These results show one representative experiment. Every point was done in triplicate. (From Garcia-Peñarrubia *et al.* (1989) (109).)

effector cells were contaminated by less than 0.5 per cent monocytes identified as non-specific esterase-positive cells. Although these cells may have contributed to the bactericidal activity observed, this contribution was most likely minimal. In any case, only two experiments using strains of *S. typhimurium* were shown in this study (30), and this bacterium is relatively resistant to NK antibacterial activity (109). This fact, together with the wide variability in antibacterial activity expressed by NK cell preparations from different donors and the increased antibacterial activity observed in overnight cultured NK cells compared to fresh isolated effector cells (109), may explain the inability of this group to identify NK cell antibacterial activity. Cell populations lacking Leu11⁺ cells did not inhibit bacterial growth. On the other hand, it should be emphasized that since T-cells were removed in the negative panning procedure (Fig. 3.1), CD3⁺, CD56⁺ cells were not examined for antibacterial activity.

5.2.2   *Effect of stimulation by staphylococcal enterotoxin B*

After activation of monocyte-depleted, Percoll-fractionated or Leu11-enriched cells with staphylococcal enterotoxin B (SEB), the same cell populations were washed and tested both for antibacterial activity against *S. typhi* (Fig. 3.4a) and cytotoxic activity against K562 (Fig. 3.4b). For each cell preparation, the antibacterial activity was significantly enhanced by SEB activation ($p < 0.05$). Similarly, the tumouricidal activity of the different effector cell preparations (Fig. 3.4b) was enhanced by SEB activation as previously discussed (85–86). In conclusion, the antibacterial activity of NK cells is increased by; (1) overnight incubation, probably through the expression of IL-2R and synthesis of IL-2, (2) SEB stimulation which is an IL-2-dependent phenomenon (85), and (3) increasing the purity of CD16$^+$, CD56$^+$ cells, since other cell populations can exert a negative control on NK activity (see Section 5.6).

5.3   Kinetics of antibacterial activity of NK cells

The kinetics of the antibacterial activity of Leu11-enriched cells, analysed as CFU (colony-forming units) of bacteria after varying incubation periods, revealed that it was dependent both on bacteria targets and the dose of

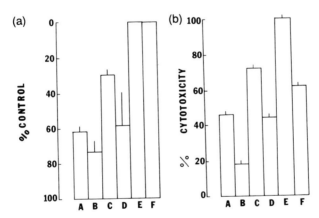

Fig. 3.4 Antibacterial activity against *S.typhi* (a) and antitumour killing against K562 cells (b) for different effector cell populations. Effector/bacteria ratio was 25/1 for both bacteria and K562 cells. Effector cells: SEB-stimulated monocyte-depleted cells (A); monocyte-depleted cells (B); SEB-Percoll fraction 1+2 (C); Percoll fraction 1+2 (D); SEB-Leu11-enriched cells (E); Leu11-enriched cells (F). SEB stimulation was performed by incubation of the cell populations with 10 μl of SEB (0.5 mg/ml), 16 h at 37 °C. The cells were washed just before they were used. Bars represent the means of 2–3 experiments ±s.d. (From Garcia-Peñarrubia *et al.* (1989) (109).)

effector cells (109). To investigate the mechanism of NK cell-mediated cytotoxicity against *E. coli* a mathematical model was developed for the kinetics of the process to which experimental data was fitted (111). The kinetics of killing were characterized by an initial bacterial target multiplication followed by rapid bacterial death. Experiments demonstrated that for each donor there is a threshold number of effector cells necessary to observe a net killing effect. Below this threshold, even at high effector to target ratios, bacterial growth cannot be stopped (111). Performing the assay at 4 °C raised the threshold number required for killing. Experiments performed in Boyden chambers confirm that NK cell–bacterial contact is not necessary for efficient killing, although bacterial lysis is slower. The close correlation between model and data supported the hypothesis that the bactericidal mechanism is extracellular (111), although the nature of the factor(s) involved in this process remains elusive.

5.4   Ultrastructural features of the interaction between NK cells and bacteria

Morphological features of the relationship between NK cells and bacteria were studied by scanning electron micrograph (SEM) and transmission electron microscopy (TEM). With TEM, NK cells enriched by panning showed the typical morphology of LGLs (109). Sections through pseudopodial regions reveal numerous electron-dense granules and vesicles. Cells often showed eccentric nuclei and cytoplasm containing prominent mitochondria, numerous free ribosomes, and a well-developed Golgi complex. In turn, rough endoplasmic reticulum is sparse. The effector cells under SEM display a uniform array of microvilli and microridges (Fig. 3.5). Some of the cells were distorted, presenting cytoplasmic expansions or pseudopodia arising from one pole of the cell, suggesting a state of activation and migratory activity (Fig. 3.6). TEM studies of the antibacterial assays after incubation in serum-free media (HB101) at 37 °C showed bacterial cell wall ghosts bound to the NK cell membrane. No bacterial survival were seen after incubation at 37 °C (109). In contrast, when the antibacterial assay was performed at 0–4 °C (which slows the bactericidal process) (111), some cells attach closely to bacteria (Figs 3.5 and 3.6), and the contact point seems to be at the tips of these extended processes of the effector cells (Figs 3.6 and 3.7). The number of bacteria surrounding NK cells was variable, ranging from none to several (Figs 3.5–3.7). In some cases, they formed rosettes around the entire NK cell surface. The variation in this feature was particularly prominent with Leu11-enriched cell isolated from a donor lacking antibacterial activity (Fig. 3.8). In this case, the surface of NK cells presented only a few microvilli and appendages, indicating a resting state (Fig. 3.8). The ability of lymphocytes to bind

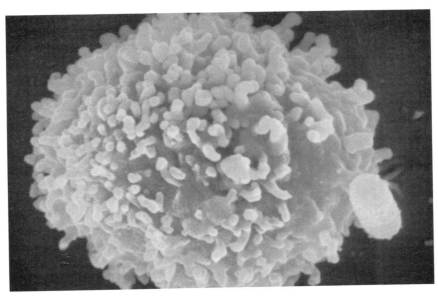

Fig. 3.5 Scanning electron micrograph of Leu11-enriched NK cells mixed with *E.coli* B/rA strain at 4°C. Intact bacterium in close association with the cytoplasmic processes on NK cells is shown.

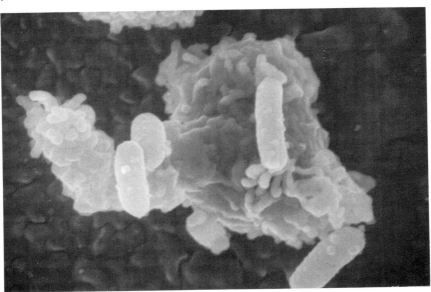

Fig. 3.6 Scanning electron micrograph of Leu11-enriched NK cells mixed with *E.coli* B/rA strain at 4°C. The surface of the NK cell shows a cytoplasmic extension or pseudopodia, and a number of bacteria are attached to the cell surface.

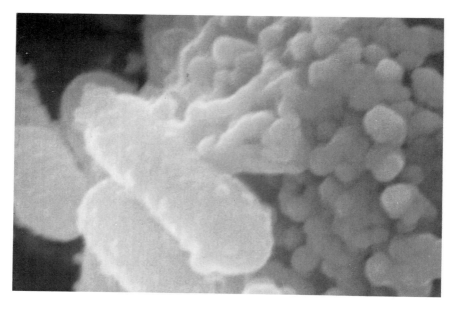

Fig. 3.7 Scanning electron micrograph of Leu11-enriched NK cells mixed with *E.coli* B/rA at 4 °C. Higher magnification of the contact points between NK cell and bacteria.

bacteria was described by Teodorescu's group several years ago (112–113). They suggested that binding of bacteria to lymphocytes may be the result of cellular mimicry. Thus, bacteria probably mimic the surface carbohydrates, or other carbohydrates present on glycoproteins and glycolipids on eukaryotic cells (112–113). However, Teodorescu and others then discovered that none of the bacteria tested were bound by NK cells. Conversely, we have found, by light microscopy (unpublished results) and EM studies (Figs 3.6–3.9), that 21.7 ± 11 per cent, 42.9 ± 7.5 per cent, and 53.4 ± 9.2 per cent of highly purified NK cells are able to bind bacteria after 15, 30, and 60 min incubation respectively. Furthermore, and in agreement with another study (32), we have observed that NK cells in the presence of serum-containing media are able to bind and phagocytose Gram+ bacteria (Fig. 3.9), but only exceptionally (less than 1 per cent of the cells) Gram− bacteria. The engulfment process is shown in Fig. 3.9 where *S. epidermidis* is seen bound to the NK cell membrane. A small microridge is being formed on the surface of the NK cell membrane to engulf the bacteria. *Staphylococcus epidermidis* is taken up by the NK cell through an invagination process, the final state of engulfment being completed when the opening of the invagination is sealed.

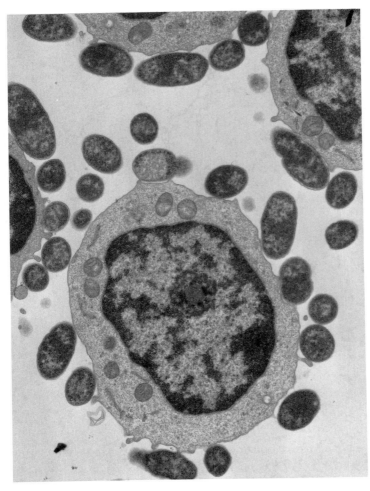

Fig. 3.8 Transmission electron micrograph of Leu11-enriched cells incubated with *S.typhimurium* at 37°C for 30 min. Intact bacteria in close association with the cytoplasmic membrane of NK cells are shown. The cells from this particular donor were lacking in antibacterial activity.

5.5   Chemotaxis of human NK cells to bacterial products

Another fact suggesting the ability of NK cells to recognize and interact with bacteria is their migratory capability to move towards the inflammation sites, and the possibility of being attracted by bacterial products. Recent studies have demonstrated that LGLs from normal human (115–116)

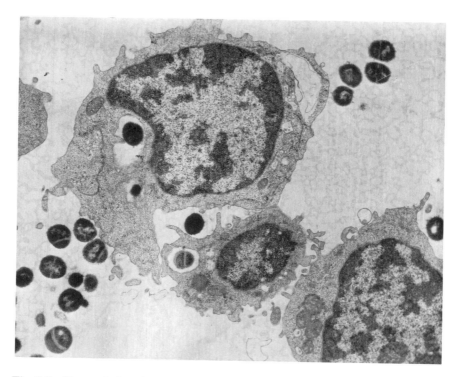

Fig. 3.9 Transmission electron micrograph of Leu11-enriched cells incubated with *S. epidermidis* at 37 °C for 15 min in RPMI supplemented with 10 per cent human AB serum. Some cells were able to phagocytose this Gram+ bacteria.

and rat (117) peripheral blood respond to chemotactic stimuli by unidirectional movement across filters in chemotactic assays performed in Boyden chambers. Active chemoattractants for normal LGLs are *N*-formyl-methionyl-leucyl-phenylalanine (f-MLP), caseine, C5a, and IFNβ. Although IL-2 was reported to have little or no effect on the migratory activity of normal human NK cells, Natuk and Welsh (118) described that IL-2 was chemotactic for mouse LGLs activated *in vivo* by virus infection, and the migration to IL-2 was inhibited by exposure to a MAB raised against the IL-2R. This indicates that IL-2 is chemotactic factor for NK cells, provided they express IL-2R. By using the leading front method of Zigmond and Hirsch (119) in Boyden chambers with 8 μm pore size nitrocellulose filters, we have performed preliminary experiments to investigate the migratory response of highly purified NK cells to bacterial products. In our system, Leu11-enriched cells were attracted by IL-2 (200 u/ml). This is consistent with the results of Natuk and Welsh (118), and could also be explained by the expression of IL-2R on Leu11-enriched cells after 16 h incubation (A. D. Bankhurst, personal communication). To study the chemotactic activity of bacterial products on NK cells we used a filtrate of *E. coli* B/rA. The filtrate was obtained by incubating bacteria in TSB (tryptic soy broth) in the exponential phase until an O.D. value of 0.35 was reached. The culture was then sterilized by passage through a 0.45 μm filter (Millipore Corp. Belford, Mass). The protein content determined in a protein assay was 2.5 μg/ml. Leu11-enriched cells (10^6) were aliquoted into the upper well of a Boyden chamber and separated by a nitrocellulose filter with 8 μm pore size from the chemoattractant placed in the lower well of the chamber. After 3 h incubation at 37°C, filters were removed and stained. Cell migration was determined by measuring the distance from the primary plane of cells to the leading front. The movement of NK cells attracted by the bacterial filtrate (88 ± 0.5 μm) was similar to that induced by IL-2 (101 ± 1 μm;, whereas control media was lacking in any chemo-attractant activity (6 ± 0.2 μm). These findings constitute further evidence that NK cells recognize and respond to bacterial antigens. However, the *in vivo* significance of this phenomenon remains to be elucidated.

5.6 Regulatory mechanism in the expression of NK cell activity

The striking increase in antibacterial activity seen between Percoll-fractionated cells (70 per cent LGLs) and panned cells (more than 95 per cent LGLs) was not proportional to the degree of enrichment in NK cells (109). This suggests that cells not bearing the Leu11/19 phenotype may be involved in the down-regulation of NK cell antibacterial activity. To test this hypothesis, we isolated Leu3(CD4$^+$) and Leu2(CD8$^+$) T-cells by nega-tive panning (Fig. 3.10). Purified CD4$^+$ or CD8$^+$ cells were co-incubated with purified autologous NK cells. No inhibitory effect was observed when NK cells were cultured overnight in the presence of CD4$^+$ cells. In contrast, the antibacterial activity of NK cells was significantly reduced after co-incubation with CD8$^+$ cells at ratios of CD8$^+$/NK cells of 1/9 and 1/4, and this effect was dramatically enhanced at ratios larger than 1/3 (Fig. 3.11a). This appears to indicate that there is a threshold number of CD8$^+$ cells necessary to achieve significant suppression of NK cell activity. In addition, the threshold number of CD8$^+$ cells could also be related to the number or activity of suppressor cells required to produce one or more suppressive factors. Cell-free supernatants (CD8-SN) were harvested from 18 h cultures of purified cytotoxic/suppressor T-cells as previously described (109). Effector cells were incubated overnight, then washed and treated with different amounts of the conditioned supernatant. The results showed that the CD8$^+$ supernatants contain one or more factors that abrogate both the antibacterial and antitumour activity (120), as well as the expression of the

NEGATIVE   PANNING

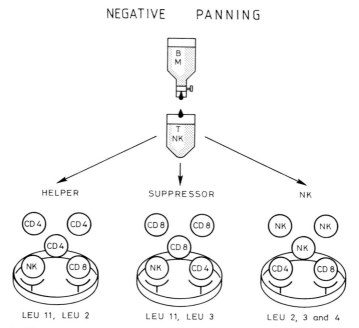

Fig. 3.10   Flow diagram of the panning procedure used to isolate purified CD4$^+$, CD8$^+$, and CD16$^+$ lymphoid cell populations. The initial steps were the same as in Fig. 3.1.

activation markers IL-2R and transferrin receptor (TR) of NK cells (data not shown). Antitumour activity of NK cells is suppressed *in vitro* by exogenous prostaglandins (121–123). To test whether prostaglandins were involved in the suppression of the NK cell antibacterial activity, NK cells were cultured overnight in the presence of different concentrations of exogenous PGE$_2$ and then tested for antibacterial activity. Figure 3.11b shows that PGE$_2$ at concentrations of 10^{-7} M completely abrogated the antibacterial activity of purified NK cells. This hypothesis was also tested by culturing the CD8$^+$ population in the presence of two different cyclooxygenase inhibitors, indomethacin, or piroxicam. Figure 3.11c shows that inhibition of prostaglandin synthesis abrogated the suppressor activity of the CD8$^+$ T-cells. These results suggest that purified cytotoxic/suppressor T-cells are able to produce prostaglandins which inhibit the activity of NK cells. We also measured the endogenous production of PGE$_2$ in overnight cultures of purified CD8$^+$ T-cells or CD8$^+$ T-cells plus indomethacin or piroxicam, without bacterial or cytokine stimulation. CD8$^+$ T-cells secreted PGE$_2$ into the culture medium in amounts (10^{-7} M) shown to inhibit NK activity when added exogenously [Fig. 3.11b and (124–125)]. Indomethacin (1 μg/ml) and piroxicam (50 mM) blocked both the PGE$_2$ synthesis and

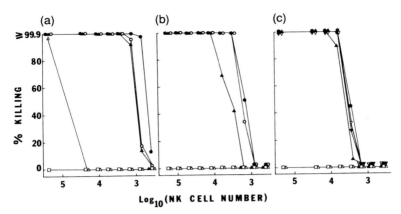

Fig. 3.11 (a) Dependence of CD8-induced suppression of the NK antibacterial activity against *E.coli* on the CD8$^+$/NK ratio. (●), NK cells control; (○), CD8$^+$ NK cells at 1/9 ratio; (▲), CD8$^+$ NK cells at 1/4 ratio; (△), CD8$^+$ NK cells at 1/3 ratio; (□), CD8$^+$ NK cells at 3/7 ratio. (b) Inhibition of the NK antibacterial activity by exogenous PGE$_2$. (●), NK cells control; (○), NK cells treated with PGE$_2$ (10^{-9}M); (▲), NK treated with PGE$_2$ (10^{-8}M); (△), NK cells treated with PGE$_2$ (10^{-7}M); (□), NK cells treated with PGE$_2$ (10^{-6}M); (c) Effect of the arachidonic acid metabolism inhibitors, indomethacin and piroxicam (Sigma Chemical Co.), on the CD8$^+$-mediated suppression of NK activity. The CD8$^+$ cell population was incubated overnight in the presence of indomethacin (1 μg/ml) or piroxicam (50 mM), washed and coincubated for 3 h with autologous NK cells at the CD8$^+$/NK ratio of 3/7 in the absence and presence of exogenous PGE$_2$. (●), NK cells control; (□), NK$^+$ CD8$^+$; (▽), NK cells plus indomethacin-treated CD8$^+$ cells; (▼), NK cells plus piroxicam-treated CD8$^+$ cells; (△), NK cells plus indomethacin-treated CD8$^+$ cells plus PGE$_2$ (10^{-7}M); (▲), NK cells plus indomethacin-treated CD8$^+$ cells plus PGE$_2$ (10^{-8}M). The data shown are representative of five identical experiments in which essentially similar results were obtained. (From Garcia-Peñarrubia *et al.* (1989) (120).)

suppression of NK cell activity (Fig. 3.11c). Again, the addition of PGE$_2$ at 10^{-7} M restored the suppression of NK antibacterial activity (Fig. 3.11c). These data demonstrate that cytotoxic/supressor T-cells (more than 98 per cent CD8$^+$ cells) produce PGE$_2$ at concentrations that suppress most T and NK cell functions *in vitro* (120–125). PGE$_2$ produced by CD8$^+$ T-cells down-regulates both the antibacterial activity and the expression of activation markers by activated NK cells to nadir levels in less than 3 h. Finally, inhibition of prostaglandin synthesis abrogates both the suppressive activity of CD8$^+$ T-cells and their PGE$_2$ production. The presence of CD8$^+$ T-cells explains why removal of relatively small numbers of T-cells in the panning procedure resulted in large increases in antibacterial and antitumour NK cell activity (109). It is notable that an *in vitro* T suppressor/NK cell ratio of 1/3 is sufficient to suppress all NK cell activity, in comparison to the higher physiological ratio for these populations in blood of aproximately 2/1. However, the *in vivo* relevance of this regulatory mechanism remains to be studied. These results are in agreement with the observations showing that i.p. injections of indomethacin at doses of 100–400 µg per mouse caused a marked activation of both NK cell activity and NK cell counts in the peritoneal cavity of mice (126). Furthermore, PGE_2 reversed the activation of NK cells induced by injection of indomethacin. In addition, a recent report of Sarin and co-workers (127) described the release of factor(s) by rat spleen cells *in vitro* that suppress the IL-2-induced NK activation of mouse spleen cells. They also showed that indomethacin could block the release of such suppressor factor(s). Likewise, they reported that IL-2 could activate rat spleen NK cells in the presence of indomethacin. These findings could explain why depletion of $CD8^+$ T-cells in PBL increased the natural antibacterial activity expressed by those populations (26, 28), as well as the absence of this antibacterial function in AIDS patients whose $CD4^+/CD8^+$ T-ratio is usually greater than 1. In addition, we have some preliminary results showing that $CD4^+$ T-cells can contra-suppress the negative effect of the $CD8^+$ T-cells when added at appropriate proportions (P. G. Peñarrubia, unpublished observations).

6   Bacterial infections in NK cell-depleted animals

From the aforementioned it can be seen that considerable evidence has accumulated to show that NK cells recognize and respond to bacteria and bacterial products *in vivo* and *in vitro*, and that NK cells are able to kill certain bacteria *in vitro*. However, the question that remains to be answered is, do NK cells play a physiological role in host's defences to bacterial infections? One approach to this is to evaluate the course of the bacterial diseases in animals selectively depleted of NK cell activity. In this context, the beige (bg) mutation in mice in its homozygous state produces a marked defect in NK activity (128). Although these animals do not seem to be highly predisposed to bacterial infections, there are some data suggesting that this may, in fact, be the case (129). Similarly, an analogous human autosomal recessive disorder, the Chediak–Higashi syndrome, is characterized by frequent pyogenic infections, with a profound NK deficiency among other clinical manifestations (130). On the other hand, the identification of asialo-GM_1 as a cell surface marker for murine NK cells has led to the production of an antibody which allows us to eliminate NK activity *in vivo* (15, 131). This procedure has been used to clarify the role of NK cells in malignant (132) and viral (133–134) diseases, and has, only recently, been applied to study the role played by NK cells in host protection against

C. trachomatis infections (50), and murine salmonellosis (135). The results of the chlamydial study showed that neither NK cell depletion by anti-GM$_1$ nor NK cell activation by the immunomodulator poly(I)poly(C) could affect quantitative tissue counts of *C. trachomatis* or survival of the mice. The results of an experimental model of murine salmonellosis have shown that injection of either virulent (SR-11) or attenuated (RIA) strain of *S. typhimurium* resulted in enhanced killing of the YAC1 targets (135). This finding is consistent with data reported by several groups with different bacterial species (Section 3). The activity peaked at 24 h after challenging, and returned to normal levels by 72 h. However, in contrast to what might be expected, mice depleted of NK cells by i.v. or i.p. injection with anti-GM$_1$ had fewer numbers of bacteria in the spleen than in undepleted animals. In the same study it was also reported that mice depleted of NK cells by anti-GM$_1$ and challenged with *L. monocytogenes* had decreased numbers of bacteria in their spleens (R. Kearns, personal communication). Based on these findings, the authors hypothesized that NK cells down-regulate the immune response when mice are challenged with a facultative intracellular bacterium.

A more selective method to deplete NK cells in mice is the administration of the NK-specific MAB, NK1.1 (136). F. T. Koster (personal communication) has studied the *in vivo* activity of murine NK cells against *S. typhimurium* infection in C57B1/6 mice. NK cells were depleted by i.p. administration of NK1.1 for 2 to 8 weeks, and the animals were challenged with either a moderately virulent strain (LT-2) or a highly virulent strain (SR-11) of the bacterium. Mice depleted of NK cells had 10-fold more LT-2 in mesenteric lymph nodes 2 days after i.p. challenge, and two-fold more LT-2, 1, 3, and 4 days after challenge. Spleens in the same mice also showed consistent increases, suggesting that NK-cell depletion temporarily permitted increased survival of *Salmonella*. However, survival of such treated mice was identical to undepleted mice at both challenge inocula and after both strains of *S. typhimurium*. According to these findings it was suggested that NK cells express antibacterial activity *in vivo* in i.p.-challenged mice, but other defence mechanisms are more important for host survival. In contrast to i.p. challenge, NK cell-depleted mice challenged orally with LT-2 survived significantly longer and this was confirmed by finding fewer *Salmonella* in Peyer's patches in NK cell-depleted mice 2 days after challenge. These findings suggest a deleterious role for NK cells at the mucosal level. This could be due to competition with IgA-armed lymphocytes at the mucosa-associated lymphoid organs (25), or to down-regulation of other cells involved in the immunoresponse, such as dendritic cells (137), or antigen-specific B- or T-cells (138–140). Whether defence against enteric infection could be enhanced by blocking the regulatory effects of intestinal NK cells merits further investigation.

7   Summary

Taken together these studies provide evidence to suggest that NK cells exert an active role in antibacterial responses. These cells are attracted to a variety of inflammatory signals, including bacterial products. They are able to migrate and concentrate at sites of inflammation. They have the ability *in vitro* to bind, phagocytose, and efficiently kill certain bacteria. They can also produce a variety of cytokines that mediate the inflammatory response by acting on other effector cells. In most cases, bacterial infections result in an early and significant increase in NK cell activity in the spleen, as well as the lymphoid compartment closer to the route of injection. Whether this is advantageous to the host's survival, either by a direct bactericidal effect, or by an indirect activation of other cell populations, remains to be elucidated. However, there is also the possibility, as suggested by some experimental *in vivo* studies, that such an elevation of NK cell activity could result in a detrimental effect on the host's defences due to down-regulation of the specific immune response by acting on antigen-presenting cells, T-cells or B-cells. Evidence has been presented supporting the hypothesis that the expression of the antibacterial activity *in vitro* is regulated by cytotoxic/suppressor T-cells through negative signals such as PGE_2, and probably contra-suppressed by helper T-cells. Once the antibacterial potential of NK cells has been established several questions remain to be answered. An important area for future research is to clarify the involvement of these cells in the pathogenesis of bacterial infections. It would be extremely helpful to understand how NK cells recognize and are activated by bacteria; which cellular interactions and immunoregulatory mechanisms are predominant in the course of different bacterial infections; and finally, whether these regulatory mechanisms can be manipulated in the therapy of established severe infections, or in the prevention of bacterial diseases in immunocompromised patients.

Acknowledgements

The original work described in this publication was performed during a three year study leave in the laboratories of Dr Frederick T. Koster and Dr Arthur D. Bankhurst of the School of Medicine at the University of New Mexico. The work was supported in part, by grants from the National Institute of Health (CA-24773 to A. D. Bankhurst, 1-R-22AI-24522 to F. T. Koster), by the UNM Research Allocation Committee (R8109), and by the Programa Sectorial de Promoción General del Conocimiento (project PM90-0042) and Comunidad Autónoma de la Región de Murcia (project PB90/18) to P. G. Peñarrubia.

References

1 Herberman, R. B. and Ortaldo, J. R. (1981). Natural killer cells: their role in defences against disease. *Science*, **214**, 24–30.

2 Lanier, L. L., Le, A. M., Civin, C. I., Loken, M. R., and Phillips, J. H. (1986). The relationship of CD16 (Leu-11) and Leu-19 (NKH-1) antigen expression on human peripheral blood NK cells and cytotoxic T lymphocytes. *J. Immunol.*, **136**, 4480–86.

3 Trinchieri, G. (1989). Biology of natural killer cells. *Adv. Immunol.*, **47**, 187–376.

4 Ritz, J., Smidt, R. E., Michon, J., Hercend, T., and Schlossman, S. T. (1988). Characterization of functional surface structures on human natural killer cells. *Adv. Immunol.*, **42**, 181–212.

5 Anderson, P., Caliguri, M., Ritz, J. and Schlossman, S. F. (1989). CD3-negative natural killer cells express ζ TcR as part of a novel molecular complex. *Nature*, **341**, 159–62.

6 Lanier, L. L., Yu, G. and Phillips, J. H. (1989). Co-association of CD3 ζ with a receptor (CD16) for IgG Fc on human natural killer cells. *Nature*, **342**, 803–5.

7 Kiesling, R., Klein, E., Proso, H., and Wigzell, H. (1975). 'Natural' killer cells in the mouse. II. Cytotoxic cells with specificity for mouse Maloney leukemia cells. Characteristics of the killer cell. *Eur. J. Immunol.*, **5**, 112–16.

8 Herberman, R. B., Nunn, M. E., and Lavrin, D. H. (1975). Natural cytotoxic reactivity of mouse lymphoid cells against syngenic and allogenic tumors. I. Distribution of reactivity and specificity. *Int. J. Cancer.*, **16**, 216–25.

9 Herberman, R. B., Reynolds, C. W., and Ortaldo, J. R. (1986). Mechanism of cytotoxicity by natural killer cells. *Ann. Rev. Immunol.*, **4**, 651–80.

10 Trinchieri, G., Masumoto-Kobayashi, M., Clark, S. C., Seehra, J., London, L., and Perussia, B. (1984). Response of resting human peripheral blood natural killer cells to interleukin 2. *J. Exp. Med.*, **160**, 1147–69.

11 Ortaldo, J. R. and Herberman, R. B. (1986). Augmentation of natural killer activity. In *Immunobiology of natural killer cells* (ed. E. Lotzová and R. B. Herberman). CRC Press, Boca Raton, Florida.

12 Yamada, S., Ruscetti, F. W., Overton, W. R., Herberman, R. B., Birchenall-Sparks, M. C., and Ortaldo, J. R. (1987). Regulation of human large granular lymphocyte and T cell growth and function by recombinant interleukin 2a: induction of interleukin 2 receptor and promotion of growth of cells with enhanced cytotoxicity. *J. Leuc. Biol.*, **41**, 505–17.

13 Ostensen, M. E., Thiele, D. L. and Lipsky, P. E. (1987). Tumor necrosis factor-α enhances cytolytic activity of human natural killer cells. *J. Immunol.*, **138**, 4185–91.

14 Anegon, I., Cuturi, M. C., Trinchieri, G., and Perussia, B. (1988). Interaction of Fc receptor (CD16) with ligand induces transcription of interleukin 2 receptor (CD25) and lymphokine genes and expression of their products in human natural killer cells. *J. Exp. Med.*, **167**, 452–72.

15 Stein-Streilein, J. (1989). Natural effector cells in influenza virus infection. In *Functions of the natural immune system* (ed. C. W. Reynolds and R. H. Witrout). Plenum, New York.

16 Lopez, C. and Fitzgerald-Bocarsly, P. (1989). Natural host defense systems active against herpes simplex virus infections. In *Functions of the natural immune system* (ed. C. W. Reynolds and A. H. Witrout). Plenum, New York.

17 Welsh, R. M., Natuk, R. J., McIntyre, K. W., Yang, H., Biron, C. A., and Bukowski, J. (1989). Factors influencing the control of virus infections by natural killer cells. In *Functions of the natural immune system* (ed. C. W. Reynolds and A. H. Witrout). Plenum, New York.

18 Albright, J. W. and Albright, J. F. (1989). Natural killer cell-mediated resistance to animal parasites. In *Functions of the natural immune system* (ed. C. W. Reynolds and A. H. Witrout). Plenum, New York.

19 Murphy, J. W. (1989). Natural host resistance mechanisms against systemic mycotic agents. In *Functions of the natural immune system* (ed. C. W. Reynolds and A. H. Witrout). Plenum, New York.

20 Reynolds, C. W. and Witrout, A. W. (1989). Summary of Part II. In *Functions of the natural immune system* (ed. C. W. Reynolds and A. H. Witrout). Plenum, New York.

21 Greenberg, A. H. (1989). Role of natural killer cells in inflammation and antibacterial activity. In *Functions of the natural immune system* (ed. C. W. Reynolds and A. H. Witrout). Plenum, New York.

22 Lowell, G. H., Smith, L. F., Artenstein, M. S., Nash, G. S., and MacDermott, R. P. (1979). Antibody-dependent cell-mediated antibacterial activity of human mononuclear cells. I. K lymphocytes and monocytes are effective against meningococci in cooperation with human immune sera. *J. Exp. Med.*, **150**, 127–37.

23 Lowell, G. H., MacDermott, R. P., Summers, P. L., Reeder, A. A., Bertovich, M. J., and Formal, S. B. (1980). Antibody-dependent cell-mediated antibacterial activity: K lymphocytes, monocytes and granulocytes are effective against *Shigella*. *J. Immunol.*, **125**, 2778–84.

24 Nencioni, L., Villa, L., Boraschi, D., and Tagliabue, A. (1983). Natural and antibody-dependent cell-mediated activity against *Salmonella typhimurium* by peripheral and intestinal lymphoid cells in mice. *J. Immunol.*, **130**, 903–7.

25 Tagliabue, A., Nencioni, L., Villa, L., Kerent, D. F., Lowell, G. H.,

and Boraschi, D. (1983). Antibody-dependent cell-mediated anti-bacterial activity of intestinal lymphocytes with secretory IgA. *Nature,* **306,** 184–6.

26 Tagliabue, A., Villa, L., Boraschi, D., Peri, G., de Gori, V., and Nencioni, L. (1985). Natural antibacterial activity against *Salmonella typhi* by human T4+ lymphocytes armed with IgA antibodies. *J. Immunol.,* **135,** 4178–82.

27 Tagliabue, A., *et al.* (1986). IgA-driven T cell-mediated antibacterial immunity in man after live oral Ty21a vaccine. *J. Immunol.,* **137,** 1504–10.

28 Tagliabue, A., *et al.* (1988). Impairment of *in vitro* natural antibacterial activity in HIV-infected patients. *J. Immunol.,* **141,** 2607–11.

29 Romano, M., *et al.* (1989). Natural antibacterial activity against *Salmonella typhi* in human cord blood. *J. Immunol.,* **142,** 2513–18.

30 Tarkkanen, J., Saxsén, H., Nurminen, M., Makela, P. H., and Saksela, E. (1986). Bacterial induction of human activated lymphocyte killing and its inhibition by lypopolysaccharide (LPS). *J. Immunol.,* **136,** 2662–9.

31 Tarkkanen, J., Saksela, E., and Lanier, L. L. (1986). Bacterial activation of human natural killer cells. Characteristics of the activation process and identification of the effector cells. *J. Immunol.,* **137,** 2428–33.

32 Abo, T., *et al.* (1986). Selective phagocytosis of Gram-positive bacteria and Interleukin 1-like factor production by a subpopulation of large granules lymphocytes. *J. Immunol.,* **136,** 3189–97.

33 Sestini, P., Nencioni, L., Villa, L., Boraschi, D., and Tagliabue, A. (1987). Antibacterial activity against *Streptococcus pneumoniae* by mouse lung lymphocytes. *Adv. Exp. Med. Biol.,* **216A,** 517–25.

34 Sestini, P., Nencioni, L., Villa, L., Boraschi, D., and Tagliabue, A. (1988). IgA-driven antibacterial activity against *Streptococcus pneumoniae* by mouse lung lymphocytes. *Am. Rev. Respir. Dis.,* **137,** 138–44.

35 Maoleekoonpairoj, S., *et al.* (1989). Lack of protection against bacterial infections in patients with advanced cancer treated by biological response modifiers. *J. Clin. Microbiol.,* **27,** 2305–8.

36 Klimpel, G. R., Niesel, D. W., and Klimpel, K. D. (1986). Natural cytotoxic effector cell activity against *Shigella flexneri*-infected HeLa cells. *J. Immunol.,* **136,** 1081–6.

37 Niesel, D. W., Hess, C. B., Cho, Y. J., Klimpel, K. D., and Klimpel, G. R. (1986). Natural and recombinant interferons inhibit epithelial cell invasion by *Shigella* spp. *Infect. Immun.,* **52,** 828–33.

38 Klimpel, G. R., Niesel, D. W., Asuncion, M., and Klimpel, K. D. (1988). Natural killer cells activation and interferon production by peripheral blood lymphocytes after exposure to bacteria. *Infect. Immun.,* **56,** 1436–41.

39 Turco, J. and Winkler, H. H. (1983). Cloned mouse interferon-gamma inhibits the growth of *Rickettsia prowazekii* in cultured mouse fibroblasts. *J. Exp. Med.,* **158,** 2159–64.

40 Hanna, L., Merigan, T., and Jawetz, J. (1966). Inhibition of TRIC agents by virus-induced interferon. *Proc. Soc. Exp. Biol. Med.,* **122,** 417–21.

41 Whitaker-Dowling, P., Dowling, J. N., Liu, L., and Youngner, J. S. (1986). Interferon inhibits the growth of *Legionella micdacei* in mouse L cells. *J. Interferon Res.,* **6,** 107–14.

42 Remington, J. S. and Merigan, T. C. (1968). Interferon protection of cells infected with intracellular protozoan *(Toxoplasma gondii). Science,* **161,** 804–6.

43 Blanchard, D. K., Stewart II, W. E., Klein, T. W., Friedman, H., and Djeu, J. Y. (1987). Cytolytic activity of peripheral blood leucocytes against *Legionella pneumophila*-infected monocytes: characterization of the effector cell and augmentation by Interleukin 2. *J. Immunol.,* **139,** 551–6.

44 Blanchard, D. K., Friedman, H., Stewart II, W. E., Klein, T. W. and Djeu, J. Y. (1988). Role of gamma interferon in induction of natural killer activity by *Legionella pneumophila in vitro* and in an experimental murine infection model. *Infect. Immun.,* **56,** 1187–93.

45 Katz, P., Yeager, Jr., H., Whalen, G., Evans, M., Swartz, R. P., and Roecklein, J. (1990). Natural killer cell-mediated lysis of *Mycobacterium-avium* Complex-infected monocytes. *J. Clin. Immunol.,* **10,** 71–7.

46 Zychlinsly, A., Karim, M., Nonacs, R., and Young, J. D-E. (1990). A homogeneous population of lymphokine-activated killer (LAK) cells is incapable of killing virus-, bacteria-, or parasite-infected macrophages. *Cell Immunol.,* **125,** 261–7.

47 Holmberg, L. A., Springer, T. A., and Ault, K. A. (1981). Natural killer activity in the peritoneal exudates of mice infected with *Listeria monocytogenes*: characterization of the natural killer cells by using monoclonal rat anti-immune macrophage antibody (M1/70). *J. Immunol.,* **127,** 1792–9.

48 Holmberg, L. A. and Ault, K. A. (1986). Characterization of *Listeria monocytogenes*-induced murine natural killer cells. *Immunol. Res.,* **5,** 50–60.

49 Kearns, R. J. and Leu, R. W. (1984). Modulation of natural killer activity in mice following infection with *Listeria monocytogenes. Cell Immunol.,* **84,** 361–71.

50 Williams, D. M., Schachter, J., and Grubbs, B. (1987). Role of natural killer cells in infection with the mouse pneumonitis agent (murine *Chlamydia trachomatis). Infect. Immun.,* **55,** 223–6.

51 Williams, D. M., *et al.* (1990). A role *in vivo* for tumor necrosis factor

alpha in host defence against *Chlamydia trachomatis. Infect. Immun.*, **58**, 1572–6.

52 Bancroft, G. J., Sheeham, K. C. F., Schrieber, R. D., and Unanue, E. R. (1989). Tumor necrosis factor is involved in the T-cell independent pathway of macrophage activation in scid mice. *J. Immunol.*, **143**, 127–30.

53 Williams, D. M., Byrne, G. I., Grubbs, B., Marshal, T. J., and Schachter, J. (1988). Role *in vivo* for gamma interferon in control of pneumonia caused by *Chlamydia trachomatis* in mice. *Infect. Immun.*, **56**, 3004–6.

54 Paya, C. V., Kenmutsu, N., Schoon, R. A., and Leibson, P. J. (1988). Tumor necrosis factor and lymphotoxin secretion by human natural killer cells leads to antiviral cytotoxicity. *J. Immunol.*, **141**, 1989–95.

55 Havell, E. A. (1989). Evidence that tumor necrosis factor has an important role in antibacterial resistance. *J. Immunol.*, **140**, 2894–9.

56 Kurlander, R. J., Hoffman, M., Kratz, S. S., and Gates, J. (1989). Comparison of the effects of IL-1α and TNF-α on phagocyte accumulation and murine antibacterial immunity. *Cell Immunol.*, **123**, 9–22.

57 Gemski, P. (1989). Pretreatment with recombinant murine tumor necrosis factor α/cachectin and murine interleukin 1α protects mice from lethal bacterial infections. *J. Exp. Med.*, **169**, 2021–7.

58 Manor, E. and Sarov, I. (1990). Inhibition of *Rickettsia conorii* growth by recombinant tumor necrosis factor alpha: enhancement of inhibition by gamma interferon. *Infect. Immun.*, **58**, 1886–90.

59 Pohajdak, B., Gomez, J. C., Wilkins, J. A., and Greenberg, A. H. (1984). Tumor-activated NK cells trigger oxidative metabolism. *J. Immunol.*, **133**, 2430–6.

60 Gomez, J., Pohajdak, B., O'Neill, S., Wilkins, J., and Greenberg, A. H. (1985). Activation of rat and human alveolar macrophage intracellular microbicidal activity by a pre-formed cytokine. *J. Immunol.*, **135**, 1194–200.

61 Djeu, J. and Blanchard, D. K. (1987). Regulation of human polymorphonuclear neutrophil (PMN) activity against *Candida albicans* by large granular lymphocytes via release of a PMN-activating factor. *J. Immunol.*, **139**, 2761–7.

62 Wolfe, S. A., Tracey, D. E., and Henney, D. S. (1976). Induction of 'natural killer' cells by BCG. *Nature*, **262**, 584–6.

63 Piccolella, E., *et al.* (1986). Effects of dexamethasone on human natural killer cell cytotoxicity, interferon production, and interleukin-2 receptor expression induced by microbial antigens. *Infect. Immun.*, **51**, 712–14.

64 Adam, A., Petit, J. F., Lefrancier, P., and Lederer, E. (1981). Muramyl peptides. Chemical structure, biological activity and mechanism of action. *Mol. Cell. Biochem.*, **41**, 27–47.

65 Goguel, A. F., Lespinats, G., and Nauciel, C. (1982). Peptidoglycan extracted from Gram-positive bacteria: expression of antitumor activity according to peptide structure and route of injection. *JNCI*, **68,** 657–63.

66 Goguel, A. F., Payelle, B., Lespinats, G., and Nauciel, C. (1983). Cytotoxic responses induced by peptidoglycans with or without *in vivo* antitumor activity. *JNCI,* **71,** 325–30.

67 Tamiyama, T. and Holden, H. T. (1979). Direct augmentation of cytolytic activity of tumor-derived macrophages in macrophage cell lines by muramyl dipeptide. *Cell Immunol.,* **48,** 369–74.

68 Rasanen, L. and Arvilommi, H. (1982). Cell walls, peptidoglycans, and teichoic acid of grampositive bacteria as polyclonal induced and immunomodulators of proliferative and lymphokine response of human B- and T-lymphocytes. *Infect. Immun.,* **35,** 523–27.

69 Sharma, S. D., Tsai, V., Krahenbuhl, J. L., and Remington, J. S. (1981). Augmentation of mouse natural killer cell activity by muramyl dipeptide and its analogs. *Cell. Immunol.,* **62,** 101–9.

70 Goguel, A. F., Payelle, B., Quan, P. C., and Lespinats, G. (1984). Stimulation of one particular subset of natural killer cells by peptidoglycans extracted from gram positive bacteria. *JNCI,* **73,** 697–703.

71 Le Garrec, Y. and Morin, A. (1987). Modulation of natural killer activity by muramyl peptides: relationships with adjuvant anti-infectious properties. *Nat. Immun. Cell. Growth Regul.,* **6,** 65–76.

72 Masihi, K. M., Lange, W., and Rohde-Schulz, B. (1987). Modulation of natural killer cytotoxicity by muramyl dipeptide and trehalose dimycolate incorporated in squalene droplets. *Cancer Immunol. Immunother.,* **24,** 19–24.

73 Gabrilovac, J., Tomasic, J., Boranic, M., Martin-Kleiner, I., and Osmak, M. (1989). *In vivo* and *in vitro* modulation of NK and ADCC activities of mouse spleen cells by peptidoglycan monomer (PGM). *Res. Exp. Med.,* **189,** 265–73.

74 Talmadge, J. E., Schneider, M., Collins, M., Phillips, H., Herberman, R. B., and Wiltrout, R. H. (1985). Augmentation of NK cell activity in tissue specific sites by liposomes incorporating MTP-PE. *J. Immunol.,* **135,** 1477–83.

75 Yokota, S., *et al.* (1988). Augmentative effect of *Nocardia rubra* cell-wall skeleton (N-CWS) on lymphokine-activated killer (LAK) cell induction. *Cancer Immunol. Immunother.,* **26,** 11–17.

76 Kawaguchi-Nagata, K., Okamura, H., Shoji, K., Kanagawa, H., Semma, M., and Shinigawa, K. (1985). Immunomodulating activities of staphylococcal enterotoxins. I. Effects on *in vivo* antibody responses and contact sensitivity reaction. *Microbiol. Immunol.,* **29,** 183–93.

77 Peavy, D. L., Adler, W. H., and Smith, R. T. (1970). The mitogenic effect of endotoxin and staphylococcal enterotoxin B on mouse spleen cells and human peripheral lymphocytes. *J. Immunol.,* **105,** 1453–8.

78 Smith, B. G. and Johnson, H. M. (1975). The effect of staphylococcal enterotoxins on the primary *in vitro* immune response. *J. Immunol.*, **115**, 575–85.

79 Zehavi-Willner, T. and Berke, G. (1986). The mitogenic activity of staphylococcal enterotoxin B (SEB): a monovalent T cell mitogen that stimulates cytolytic T lymphocytes but cannot mediate their lytic interaction. *J. Immunol.*, **137**, 2682–7.

80 White, J., Herman, A., Pullen, A. M., Kubo, R., Kapler, J. W., and Marrack, P. (1989). The V$_\beta$-specific superantigen staphylococcal enterotoxin B: Stimulation of mature T cells and clonal deletion in neonatal mice. *Cell*, **56**, 27–35.

81 Lee, C. L. Y., Lee, S. H. S., Jay, F. T., and Rozee, K. R. (1990). Immunobiological study of interferon-gamma-producing cells after staphyloccocal enterotoxin B stimulation. *Immunology*, **70**, 94–9.

82 Vroegop, S. M. and Buxser, S. E. (1989). Cell surface molecules involved in early events in T-cell mitogenic stimulation by staphylococcal enterotoxins. *Infect. Immun.*, **57**, 1816–24.

83 Herrmann, T., Accolla, R. S., and MacDonald, H. R. (1989). Different staphylococcal enterotoxins bind preferential to distinct major histocompatibility complex class II isotypes. *Eur. J. Immunol.*, **19**, 2171–4.

84 Fleischer, B. and Schrezenmeier, H. (1988). T cell stimulation by staphylococcal enterotoxins. Clonally variable response and requirement for major histocompatibility complex class II molecules on accessory or target cells. *J. Exp. Med.*, **167**, 1697–1707.

85 Bankhurst, A. D. and Imir, T. (1989). The mechanisms involved in the activation of human natural killer cells by staphyloccocal enterotoxin B. *Cell Immunol.*, **122**, 108–21.

86 Garcia-Peñarrubia, P., Lennon, M. P., Koster, F. T., Kelley, R. O., and Bankhurst, A. D. (1989). Selective proliferation of natural killer cells among monocyte-depleted peripheral blood mononuclear cells as a result of stimulation with staphylococcal enterotoxin B. *Infect. Immun.*, **57**, 2057–65.

87 Vogel, S. N., Hilfiker, M. L., and Caulfield, M. J. (1983). Endotoxin induced T lymphocytes proliferation. *J. Immunol.*, **130**, 1774–9.

88 Rietschel, E. T., Schade, U., Jensen, M., Wollenweber, H. W. Luderitz, O., and Greisman, S. G. (1982). Bacterial endotoxins: chemical structure, biological activity and role in septicaemia. *Scand. J. Infect. Dis.* (Suppl.), **31**, 8–21.

89 Morrison, D. C. and Rudbach, J. A. (1981). Endotoxin-cell membrane interactions leading to transmembrane signaling. *Contemp. Top. Mol. Immunol.*, **8**, 187–218.

90 Miller, R. A., Gartner, S., and Kaplan, H. S. (1978). Stimulation of mitogenic responses in human peripheral blood lymphocytes by lipo-

polysaccharide: serum and T helper cell requirements. *J. Immunol.*, **121**, 2160–4.

91 Ohta, M., Rothmann, J., Kovats, E., Pham, P. H., and Nowotny, A. (1985). Biological activities of lipopolysaccharide fractionated by preparative acrylamide gel electrophoresis. *Microbiol. Immunol.*, **29**, 1–12.

92 Lindemann, R. A., Miyasaki, K. T., and Wolinsky, L. E. (1988). Induction of activated lymphocyte killing by bacteria associated with periodontal disease. *J. Dent. Res.*, **67**, 846–850.

93 Lindemann, R. A. (1988). Bacterial activation of human natural killer cells: role of cell surface lipopolysacharide. *Infect. Immun.*, **56**, 1301–8.

94 Lindemann, R. A. (1989). Role of interferon and cellular adhesion molecules in bacterial activation of human natural killer cells. *Infect. Immun.*, **57**, 1702–6.

95 Salata, R. A., Kleinherz, M. E., Schacter, B. Z., and Ellner, J. J. (1984). Augmentation of natural killer cell activity by lipopolysaccharide through separable effects on the binding of non-adherent lymphocytes to tumor targets and tumor killing. *Cancer Res.*, **44**, 1044–7.

96 Kang, Y. H., Carl, M., Maheshwari, R. K., Watson, L. P., Yaffe, L., and Grimley, P. M. (1988). Incorporation of bacterial lypopolysaccharide by human Leu-11 a$^+$ natural killer cells: ultrastructural and functional correlations. *Lab. Invest.*, **58**, 196–209.

97 Kang, Y. H. (1990). Effects of bacterial lypopolysaccharide and calmodulin on Ca^{2+}-ATPase and calcium in human natural killer cells, studied by a combined technique of immunoelectron microscopy and ultracytochemistry. *J. Histochem. Cytochem.*, **38**, 359–370.

98 Klein, J. S. and Kearns, R. J. (1989). Analysis of the kinetics of natural killer cell activity in mice during an active infection with *Pseudomonas aeruginosa*. *Nat. Immun. Cell Growth Regul.*, **8**, 37–47.

99 Lai, W. C., *et al.* (1990). Resistance to *Mycoplasma pulmonis* mediated by activated natural killer cells. *J. Infect. Dis.*, **161**, 1269–75.

100 Schwab, J. H. (1975). Suppression of the immune response by microorganisms. *Bacteriol. Rev.*, **39**, 121–43.

101 Reimann, J., Claesson, M. H., and Qvirin, N. (1990). Suppression of the immune response by microorganisms. *Scand. J. Immunol.*, **31**, 543–6.

102 Ito, M., Ralph, P., and Moore, M. A. S. (1980). Suppression of spleen natural killing activity induced by BCG. *Clin. Immunol. Immunopathol.*, **16**, 30–8.

103 Savary, C. A. and Lotzová, E. (1978). Suppression of natural killer cytotoxicity by splenocytes from *Corynebacterium parvum*-infected bone marrow-tolerant and infant mice. *J. Immunol.*, **120**, 239–43.

104 Lotzová, E. and Savary, C. A. (1986). Regulation of NK cell activity

by suppressor cells. In *Immunobiology of natural killer cells,* Vol. II (ed. E. Lotzová and R. B. Herberman). CRC Press, Boca Raton, Florida.

105 Savary, C. A. and Lotzová, E. (1987). Mechanism of decline of natural killer cell activity in *Corynebacterium parvum*-treated mice: inhibition by erythroblasts and Thy 1.2^+ lymphocytes. *JNCI,* **79,** 533–41.

106 Zighelboim, J., Lichtenstein, A., Bich, A., and Mickel, R. (1983). Inhibition of natural killer cell activity by a soluble substance released by rat peritoneal cells. *Cancer Res.,* **43,** 1984–9.

107 Pedersen, B. K. and Kharazmi, A. (1987). Inhibition of human natural killer cell activity by *Pseudomonas aeruginosa* alkaline protease and elastase. *Infect. Immun.,* **55,** 986–9.

108 Rechnitzer, C., Diamant, M., and Pedersen, B. K. (1989). Inhibition of human natural killer cell activity by *Legionella pneumophila* protease. *Eur. J. Clin. Microbiol. Infect. Dis.,* **8,** 989–92.

109 Garcia-Peñarrubia, P., Koster, F. T., Kelley, R. O., McDowell, T. D., and Bankhurst, A. D. (1989). Antibacterial activity of human natural killer cells. *J. Exp. Med.,* **169,** 99–113.

110 Timonen, T. and Saksela, E. (1980). Isolation of human natural killer cells by density gradient centrifugation. *J. Immunol. Methods,* **36,** 285–91.

111 Garcia-Peñarrubia, P., Bankhurst, A. D., and Koster, F. T. (1989). Experimental and theoretical kinetics study of antibacterial killing mediated by human natural killer cells. *J. Immunol.,* **142,** 1310–17.

112 Teodorescu, M. and Mayer, E. P. (1982). Binding of bacteria to lymphocyte subpopulations. *Adv. Immunol.,* **33,** 307–51.

113 Teodorescu, M. (1984). Binding of bacteria in lymphocyte subpopulations: role of lectin-carbohydrate interactions. *Biol. Cell.,* **51,** 251–8.

114 Antonaci, S., Jirillo, E., Ventura, M. T., Michalek, S. M., Bonomo, L., and McGhee, J. R. (1984). Relationship between immune system and gram-negative bacteria. II. Natural killer cytotoxicity of *Salmonella-minnesota* R345-unbound peripheral blood lymphocytes. *J. Immunol.,* **133,** 729–33.

115 Botazzi, B., Introna, M., Allavena, P., Villa, A., and Mantovani, A. (1985). *In vitro* migration of human large granular lymphocytes. *J. Immunol.,* **134,** 2316–21.

116 Pohajdak, B., Gomez, J., Orr, F. W., Khalil, N., Talgoy, M., and Greenberg, A. H. (1986). Chemotaxis of large granular lymphocytes. *J. Immunol.,* **136,** 278–84.

117 Punturieri, A., Santoni, A., Ming, W. J., Nobili, N., Mantovani, A., and Botazzi, B. (1989). *In vitro* migration of rat large granular lymphocytes. *Cell Immunol.,* **123,** 257–63.

118 Natuk, R. J. and Welsh, R. M. (1987). Chemotactic effect of human recombinant interleukin-2 on mouse activated large granular lymphocytes. *J. Immunol.*, **139**, 2737–43.

119 Zigmond, S. H. and Hirsch, J. G. (1973). Leukocyte locomotion and chemotaxis: new methods for evaluation, and demonstration of a cell-derived chemotactic factor. *J. Exp. Med.*, **137**, 387–410.

120 Garcia-Peñarrubia, P., Bankhurst, A. D., and Koster, F. T. (1989). Prostaglandins from human T suppressor/cytotoxic cells modulate natural killer antibacterial activity. *J. Exp. Med.*, **170**, 601–6.

121 Lang, N., Ortaldo, J. R., Bonnard, G. D., and Herberman, R. B. (1982). Interferon and prostaglandins: effects on human natural and lectin-induced cytotoxicity. *JNCI*, **69**, 339–43.

122 Leung, K. H. and Koren, H. S. (1982). Regulation of human natural killing. II. Protective effect of interferon on NK cells from suppression by PGE$_2$. *J. Immunol.*, **129**, 1742–7.

123 Leung, K. H. (1989). Inhibition of human NK cell and LAK cell cytotoxicity and differentiation by PGE$_2$. *Cell Immunol.*, **123**, 384–95.

124 Goodwin, J. S. and Ceuppens, J. (1983). Regulation of the immune response by prostaglandins. *J. Clin. Immunol.*, **3**, 295–315.

125 Hall, T. J., Chen, S. H., Brostoff, J. and Lydyard, P. M. (1983). Modulation of human natural killer cell activity by pharmacological mediators. *Clin. Exp. Immunol.*, **54**, 493–500.

126 Voth, R., Chmielarczyk, W., Storch, E., and Kirchner, H. (1986). Induction of natural killer cell activity in mice by injection of indomethacin. *Nat. Immun. Cell Growth Regul.*, **5**, 317–24.

127 Sarin, A., Adler, W. H., and Saxena, R. K. (1989). Lack of optimal activation of natural killer levels by interleukin-2 in rat spleen cells: Evidence for suppression. *Cell Immunol.*, **122**, 548–54.

128 Roder, J. and Duive, A. (1979). The beige mutation in the mouse selectively impairs natural killer cell function. *Nature*, **278**, 451–3.

129 Morgan, D. R., Dupont, H. L., Wood, L. V., and Kohk, S. (1984). Cytotoxicity of leukocytes from normal and *Shigella*-susceptible (opium-treated) guinea pigs against virulent *Shigella sonnei*. *Infect. Immun.*, **46**, 22–4.

130 Barak, Y. and Nir, E. (1987). Chediak–Higashi syndrome. *Am. J. Pediatr. Hematol. Oncol.*, **9**, 42–55.

131 Young, W. W. Jr., Hakomori, S. I., Durdrik, J. M., and Henney, C. S. (1980). Identification of ganglio-N-tetraosylceramide as a new cell surface marker for murine natural killer (NK) cells. *J. Immunol.*, **124**, 199–201.

132 Kawase, I., Urdal, D. L., Brooks, C. G., and Henney, C. S. (1982). Selective depletion of NK cell activity *in vivo* and its effect on the growth of NK-sensitive and NK-resistant tumor cell variants. *Int. J. Cancer*, **29**, 567–74.

133 Bukowski, J. F., Woda, B. A., Habu, S., Okumura, K., and Welsh, R. M. (1983). Natural killer cell depletion enhances viral synthesis and viral-influenced hepatitis *in vivo*. *J. Immunol.*, **131**, 1531–8.

134 Stein-Streilein, J. and Gufee, J. (1986). *In vivo* treatment of mice and hamsters with antibodies to asialo GM_1 increases mortality to pulmonary influenza infection. *J. Immunol.*, **136**, 1435–41.

135 Smith, R. A., Brzezicki, M. J., Griggs, N., and Mahrer, S. (1989). The role of natural killer cells in experimental murine salmonellosis. *Nat. Immun. Cell Growth Regul.*, **8**, 331–7.

136 Seaman, W. E., Sleisenger, M., Eriksson, E., and Koo, G. (1987). Depletion of natural killer cells in mice by monoclonal antibody to NK-1.1. *J. Immunol.*, **138**, 4539–44.

137 Shah, P. D. (1987). Dendritic cells but not macrophages are target for immune regulation by natural killer cells. *Cell Immunol.*, **104**, 440–5.

138 Robles, C. P. and Pollack, S. B. (1989). Asialo-GM_1^+ natural killer cells directly suppress antibody-producing B cells. *Nat. Immun. Cell Growth Regul.*, **8**, 209–22.

139 Katz, P., Mitchell, S. R., Cupps, T. R., Evans, M., and Whalen, G. (1989). Suppression of B cell responses by a natural killer cells is mediated through direct effects on T cells. *Cell Immunol.*, **119**, 130–42.

140 Commes, T., Clofent, G., Jourdan, M., Bataille, R., and Klein, B. (1990). Human natural killer cells suppress the proliferation of B cells. *Immunol. Letts.*, **24**, 57–62.

4 Natural killer cells in viral infection

R. M. WELSH and M. VARGAS-CORTES

1   Introduction

A primary role for NK cells is to rid the body of unwanted cells with growth properties and surface structures unsuitable for normal cell function. NK cells may be of major significance in viral infections because of the ability of viruses to alter cell metabolism and architecture, and also because viruses are potent inducers of interferons (IFNs), which augment NK cell cytotoxicity and proliferation. As outlined in Chapter 1, prototypical NK cells are characterized as being large granular lymphocytes (LGLs) which mediate non-major histocompability complex (MHC)-restricted cytotoxicity without apparent specificity or antigenic memory (1). They lack T-cell receptors (TcR) and the TcR-associated CD3 molecules, but express ill-defined putative 'NK receptors' and the well-defined (CD16) Fc receptor (FcR) for immunoglobulin G (IgG). The expression of these two types of receptors allows them to mediate a non-specific 'natural killer' function as well as specific antibody-dependent cellular cytotoxicity (ADDC) against target cells coated with antibody. Both of these events are likely to play roles in virus infections. In addition to the prototypic CD3$^-$ NK cells, CD3$^+$, TcR$^+$T-cells can sometimes mediate a similar form of antigen-non-specific cytotoxicity, particularly after activation with interleukin-2 (IL-2). These cells are referred to as 'non-MHC-restricted' cytotoxic T-lymphocytes (CTLs), and their roles in immune system functions have yet to be fully clarified (1) This review will focus on the augmentation of the NK cell response, which is a feature of nearly every viral infection in mouse and man which has been examined (2), and the role of NK cells in regulating viral infections.

2   NK cell response to viral infection

The kinetics of NK cell activation during infection have been most systematically analysed in the mouse, and the CTL response to lymphocytic choriomeningitis virus (LCMV) – typical of many viruses – is illustrated in Fig. 4.1. The NK cell response becomes augmented at early stages of infection, in parallel with the induction of type 1 IFN (α/β) (4–6). This is followed by the virus-specific MHC-restricted CTL response. The augmented NK cell response that occurs after viral infection is the result of two phenomena, (1) the activation of NK cells to a higher state of cytotoxicitiy, and (2) the proliferation of NK cells, which results in an increase in NK cell number.

2.1   Activation

NK cells can be activated to a highly cytotoxic state by a variety of cytokines, most notably IL-2 and IFN α, β, and γ (4–8). A variety of viral

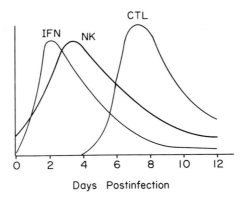

Days   Postinfection

Fig. 4.1 Cytotoxic cell response to virus infection in mice. This chart depicts the IFN α/β, NK cell, and virus-specific CTL responses in the spleen after an inocula-tion of mice with LCMV (2). (Reprinted with the permission of Clinical Immunology Newsletter, Elsevier, New York.)

glycoproteins, including those isolated from measles (9), mumps (10, 11), influenza (12), LCM (9), and Sendai (13) viruses, also have been shown to augment NK cell-mediated lysis. The major systemic activating agent for NK cells induced during a virus infection *in vivo*, however, appears to be type 1 IFN (α, β). The level of NK cell activation during infection corre-lates with systemic levels of type 1 IFN (4). Purified or recombinant IFNα or β activates NK cells *in vivo* (4–6), and antiserum to type 1 IFN blocks virus-induced activation of mouse NK cells (6). The IFN response *in vivo* wanes as virus is cleared by the developing specific T-cell response, and, concomitantly, NK cells return to their pre-infection level of activation (4, 5; Fig. 4.1). Athymic nude mice and mice with severe combined immuno-deficiency (scid) fail to generate a T-cell response or clear virus infections. As a result, IFN levels remain elevated and NK cells are activated for a prolonged time period (4, 14).

2.2   Proliferation

The proliferation of NK cells contributes greatly to the augmented NK cell response during infection. Endogenous NK cells have a relatively low turnover rate (15, 16), but a viral infection results in a profound prolifera-tive response. During viral infection, at the peak of NK cell activation, NK cells become large blast-sized cells with enhanced expression of Thy1, FcR, and nylon wool-mediated adherence (17, 18). Single cell cytotoxicity assays using effector cells pulsed for 1 h and ^{3}H-thymidine showed that little (less than 3 per cent) killing by endogenous NK cells was mediated by ^{3}H-thymidine-incorporating cells, whereas 20–30 per cent of the killing by activated NK cells 3 days after infection was mediated by the radiolabelled cells (19). Treatment of mice with a regimen of hydroxyurea which kills dividing cells had little effect on endogenous NK cells but greatly reduced the levels of activated NK cells (16). These experiments indicate that a relatively 'resting' NK cell population becomes stimulated during viral infection to proliferate rapidly.

The activation and proliferation of NK cells are closely linked, as both peak early in infection, in parallel with the IFNα/β response (4, 19). Indeed, injection of mice with purified IFNβ stimulates NK cell blasto-genesis (20). IFN is normally considered to be growth inhibitory, rather than mitogenic (21), suggesting that proliferating non-cytolytically active 'pre-NK cells' are converted by IFN into a cytolytic state. Several experiments argue against this, however. First, short-term (1 h) exposure of mice *in vivo* to IFN activates NK cells but does not elicit the appearance of blast NK cells, which requires 16–24 h *in vivo* treatment (20). Secondly, hydroxyurea greatly reduces the NK cell response in virus-infected or polyI:C-stimulated mice but does not inhibit the short-term IFN-activation mechanism (16). Thirdly, the increase in the number of LGLs expressing NK cell marker antigens in virus-infected organs may be too high to be explained without an enhanced NK cell proliferation mechanism (22, 23). Therefore, it appears that IFN directly or indirectly stimulates the prolifer-ation of NK cells.

2.3   Accumulation of NK cells in infected tissue

The accumulation of NK cells in virus infected tissue is quite impressive at the early stages of infection. Augmented *activity* of NK cells has been documented for the spleen, peripheral blood, peritoneal cavity, liver, lung, gut, and cerebrospinal fluid (CSF) following infection with viruses (re-viewed in 2, 24). The site of the activity is dependent on the injection route and the virus. For example, infection with the influenza virus via intratracheal (i.t.) inoculation leads to an activation of NK cells in the lung (25). Intracerebral inoculation with Sindbis or vaccinia virus stimulates an NK cell response in the CSF (26, 27). Increases in the *number* of NK cells (i.e. asialo-GM_1^+, $CD8^-$ LGLs) have also been documented for many organs in several viral systems (22, 23). In general, the accumulation of NK cells is thought to be a function of where the virus is replicating and the cytopathic nature of the infection. An i.p. infection of mice with mouse hepatitis virus results in a potent 150-fold increase in the number of NK cells in the peritoneal cavity (Fig. 4.2), whereas an i.v. infection, which results in a high level of virus replication in the liver, is associated with a profound leucocyte infiltrate into the liver in which up to 20 per cent of the leucocytes and 50 per cent of lymphocytes are asialo-GM_1^+, $CD8^-$ LGLs. Blast NK cells can be found in both lymphoid (e.g. spleen) and non-

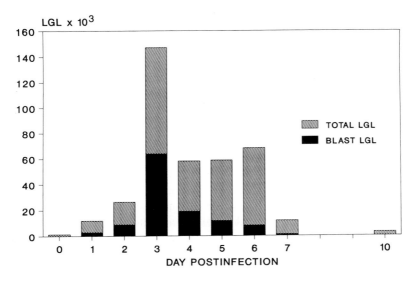

Fig. 4.2 Infiltration of NK cells in the peritoneal cavity of mice inoculated i.p. with MHV. This figure depicts the total number of LGLs and the blast LGLs, as distinguished morphologically by size, dark staining of the cytoplasm, prominent nucleoli, and/or mitotic figures. As MHV is a virus which does not induce a potent T-cell response, there is no significant LGL influx at day 7. (This figure is based on data derived from (22, 28).)

lymphoid (e.g. liver, peritoneal cavity) tissue at early stages of infection, and the level of NK cell blastogenesis in these organs correlates with IFN response (22, 23, 28). Whereas this suggests that NK cells proliferate at sites of infection, it is not known to what extent this replication takes place. Accumulation of LGLs in the liver of mice treated with the biological response modifier, MVE-2, is blocked if mice are treated with the bone marrow-seeking isotope strontium 90 (29). This indicates that bone marrow is required for the accumulation of NK cells, but does not distinguish between the bone marrow being the source of liver NK cells or providing factors for maturation and proliferation of existing NK cells in the liver. Blast NK cells in virus-infected mice are found in the spleen, but if mice are whole-body irradiated (under conditions in which the spleen is shielded), the spleen is not a sufficient source of NK cells to seed the liver or peritoneal cavity (28). More work is needed to clarify this issue.

2.4 Chemotaxis of NK cells

The accumulation of NK cells in virus-infected organs is likely to involve chemotaxis in addition to proliferation. Resting endogenous NK cells isolated from unstimulated mice respond poorly to chemotactic agents, but activated NK cells from virus-infected mice respond chemotactically to a variety of agents, including IFNβ, IL-2, and peritoneal exudates from virus-infected mice (22, 30). NK cells exhibiting blast cell morphology are preferably stimulated in these chemotaxis assays.

2.5   Assessment of the roles of IL-2 and other cytokines during the virus-induced NK cell response

The NK cell responses of activation, proliferation, and accumulation are undoubtedly influenced by a variety of cytokines in addition to IFNα/β. A case in point is that IFNα/β does not stimulate NK cell blastogenesis *in vitro*, suggesting that another factor(s) is involved in IFNα/β induction NK cell blastogenesis *in vivo*. IFNα/β might stimulate the production of an NK cell growth factor, or act synergistically with a growth factor, perhaps by inducing the appropriate receptors. Possible additional cytokines which may be involved are tumour necrosis factor alpha (TNFα), which is largely induced by the same factors which induce IFNβ (31) and whose synthesis during LCMV infection parallels the IFNα/β response (R. M. Welsh and N. Ruddle, unpublished). IFNγ is produced by NK cells activated during viral infection (32, 33) and may play some regulatory role by functioning as an autocrine factor; injection of recombinant IFNγ into mice stimulates NK cell blastogenesis (20). A novel B-cell lymphoma-produced cytokine which stimulates human NK cell proliferation *in vitro* has recently been described (34), but it is not yet known whether such a cytokine is involved in the virus-induced NK cell proliferation.

NK cells have receptors for IL-2 and can be influenced in many ways by this cytokine. IL-2 enhances NK cell cytotoxicity and, unlike IFN, IL-2 (at high concentrations) stimulates NK cell division *in vitro* (8, 35). Treatment of mice with high doses of IL-2 also stimulates NK cell activation and blastogenesis *in vivo* (36). IL-2 enhances the repopulation of mice with NK cells after bone marrow reconstitution (37) and stimulates NK cell differentiation in bone marrow cultures (38). Transgenic mice constitutively expressing IL-2 display aberrantly high numbers of NK cells (39). IL-2 also stimulates the chemotaxis of virus-induced activated NK cells *in vitro* (30). These data suggest a strong role for IL-2 in NK cell development, including the stimulation of NK cell bone marrow precursors, and the activation, proliferation, chemotaxis, and accumulation of mature NK cells.

However, the above studies for the most part have used very high concentrations of IL-2, and a number of experiments in viral systems have challenged the assumption that IL-2 is a major regulator of NK cells *in vivo*. First, NK cell blastogenesis, which peaks at 3 days post-infection in the LCMV system, does not correlate with the IL-2 response, which peaks with the T-cell response at 7–9 days post-infection (19, 28, 40). One could argue that the T-cells have IL-2 receptors with higher affinity than those on NK cells and that as the clones of specific T-cells expand they compete with NK cells for the IL-2. However, treatment of mice with cyclosporin A, which blocks IL-2 synthesis, selectively blocks T-cell proliferation but not NK cell proliferation (41). Furthermore, NK cell activation and blastogenesis proceed normally in nude mice and scid mice, both of which produce very little IL-2 (14, 19). NK cells found in lesions rich in T-cells are for the most part neither activated nor proliferating (4, 19, 28). On the other hand, NK cells do accumulate at high levels in T-cell-dependent infiltrates in the liver and peritoneal cavity of virus-infected mice. This accumulation does not occur in nude mice (23). Arguably, T-cell products, which could include IL-2 and IFNγ, or factors from tissue destroyed by T-cells, may stimulate the migration of NK cells into such tissues (30). The accumulation of NK cells is the only NK cell property which strongly correlates *in vivo* with the IL-2 response during the LCMV infection of mice.

3   Model systems examining the control of viral infections by NK cells

The mouse provides a useful model to determine the relative roles of the different elements of the host response in controlling infection. The standard techniques of selective depletion and reconstitution have revealed that NK cells play a major role in some infections but not in others. We shall initially restrict our discussion to two well-studied murine viruses, murine cytomegalovirus (MCMV) and LCMV, which have markedly different sensitivities to NK cells *in vivo*.

3.1   Murine cytomegalovirus

Perhaps the most convincing evidence that NK cells play a role in viral infection has resulted from studies of MCMV. This pantropic virus infects leucocytes, hepatocytes, and many other cell types and grows particularly well in the spleen and liver. Under conditions of high viral dose or immunosuppression, MCMV, like human CMV, disseminates to the lungs. After the mouse recovers from the acute infection, virus can be recovered from the salivary gland for 2 to 3 months, as the mice remain persistently infected (42). A suggestion that NK cells control MCMV infection comes from the observation that suckling mice, which have low NK cell activity, are very sensitive to MCMV, and that resistance develops at the third to fourth week of age, in concordance with the development of the NK cell response (43, 44). Correlations were noted between the genetically predetermined NK cell responses of mice and resistance to MCMV (45), and homozygous beige mice, which have a granule-associated NK cell deficit (46), were more sensitive than their normal heterozygous litter mates (Fig. 4.3). Resistance of beige mice to infection could be conferred by inoculation with normal bone marrow cells, which are progenitors for NK cells (47).

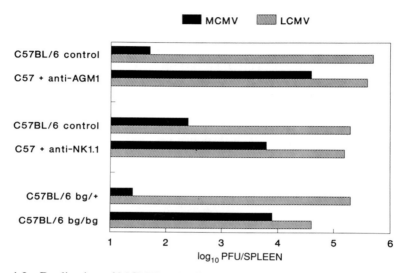

Fig. 4.3 Replication of MCMV and LCMV in mice lacking NK cell activity. This figure depicts the plaque-forming units (PFU) of MCMV and LCMV in the spleens of C57BL/6 mice depleted of NK cells with antiserum to asialo-GM_1 or antibody to NK1.1 and of NK cell-deficient beige mice. Mice were inoculated i.p. and examined for PFU at 3 days post-infection. (This figure is based on data published in (48, 49, 51 and 56).)

Selective depletion and reconstitution studies have provided more definitive evidence that NK cells regulate MCMV. Mice selectively depleted of NK cell activity by injections with antiserum to asialo-GM1 (48–51) or monoclonal antibody to NK1.1 (51, 52) developed increased sensitivity to MCMV, as judged by increases in morbidity, mortality, and titres of virus in the spleen and liver, which are major organs for MCMV replication (Fig. 4.3). Under these conditions of NK cell depletion the virus disseminated to and grew in the lung (49). The biological response modifier OK432 both enhanced NK cell activity and the resistance of mice to MCMV, whereas depletion of NK cell activity with antiserum to asialo-GM1 blocked the ability of OK432 to enhance resistance to MCMV (50).

Adoptive transfer of non-immune adult mouse splenocytes into 4–6 day old suckling mice protected the recipients from lethal MCMV infections and greatly reduced spleen MCMV plaque-forming units (PFU) early in infection (53) (Fig. 4.4). Depletion of NK cells from donor spleens abolished the protective effect, whereas depletion of T-cells, B-cells, macrophages, granulocytes, and (probably, in retrospect) dendritic cells did not inhibit this phenomenon (53). Partial purification of NK cells through nylon wool or Percoll gradients enriched for the protective effect. These experiments conclusively indicated that NK cells are an important component of natural resistance to MCMV.

Depletion of NK cell activity in athymic nude mice (53) or in scid mice (14) caused enhanced severity of MCMV infection, indicating that neither T- nor B-cell function is required for the antiviral effects of NK cells. It has been shown with human NK cells that varying levels of 'natural killing' activity against a variety of tumour and virus-infected targets may actually be forms of ADCC incuded by cytophilic antibody bound to the NK cell Fc receptors (54, 55). The experiment with scid mice indicates that, at least in the case of MCMV, the antiviral effect is not by an ADCC mechanism.

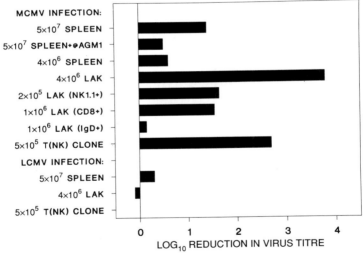

Fig. 4.4   Adoptive transfer of leucocytes into suckling mice infected with MCMV or LCMV. Four- to six-day-old suckling mice were inoculated with adult spleen leucocytes from normal and NK cell-depleted (with antiserum to asialo-GM_1) mice, with lymphokine-activated killer (LAK) cells, or with a non-MHC-restricted cytotoxic lymphocyte (CTL) clone with NK-like cytotoxicity. Mice were then challenged i.p. with MCMV or LCMV, and PFU in the spleen were determined 3 days post-infection. PFU/spleen values in the untreated controls ranged from 3.8 to 4.8 $\log_{10}$, and the data are presented as the $\log_{10}$ reduction in titre. (This figure is a compilation of several experiments (53, 60).)

Furthermore, the use of scid mice rules out antiviral effects which may be mediated by non-MHC restricted CTL which might express asialo-GM1 or NK1.1 antigens.

3.2   Lymphocytic choriomeningitis

Much work with LCMV has indicated it to be resistant to NK cells in mice (Fig. 4.3). LCMV synthesis is not significantly elevated in beige mice (56), normal mice treated with either anti-asialo-GM1 (48) or anti-NK1.1 (51), or in scid mice treated with anti-asialo-GM1 (14). In addition, it is not elevated during the first 3 days of infection of mice more globally immuno-suppressed with cyclophosphamide (57). There is no pronounced age-dependent restriction in LCMV replication that correlates with NK cell levels, and adoptive transfer of NK cell-containing adult splenocytes into suckling mice does not confer protection to LCMV infection (53) (Fig. 4.4). Depletion of NK cells in mice congenitally infected with LCMV and with lifelong persistent infections did not increase the levels of LCMV PFU/ml in the blood (58). The evidence that NK cells may control LCMV by an ADCC mechanism is also negative, as the protective effect of passively transferred immune serum was not altered by treatment with antiserum to asialo-GM1 or cyclophosphamide (57).

4   Adoptive immunoprophylaxis and therapy with lymphokine-activated killer cells

The pronounced sensitivity of MCMV to NK cells, coupled with the profound significance of human cytomegalovirus (HCMV) pneumonia as being a leading cause of death in transplant recipients and acquired im-mune deficiency syndrome (AIDS) patients, led us to explore the possibil-ity of using cultured NK cells to control MCMV infections. The term 'LAK' has been used to describe killer cells generated *in vitro* from peripheral blood mononuclear cells in the presence of IL-2 (see Chapters 1 and 6 of this volume). NK, T-, and B-cells all expand in these cultures (60). Unfractionated LAK cell populations profoundly reduced MCMV titres in prophylactically treated suckling mice, with as few as 5×10^5 cells causing a 1000-fold reduction in MCMV titres by 3 days post-infection (Fig. 4.4). The LAK cell culture was fractionated by flow cytometry into the three classes of cells, and potent protective effects were shown with NK-LAK (NK1.1$^+$) and T-LAK (CD8$^+$), but not with B-LAK (IgD$^+$) cells. This reflected the ability of these cells to kill tumour targets in culture, as the NK-LAK and T-LAK (but not B-LAK) cells lysed a variety of cultured target cells (60). Of interest was the ability of the T-LAK cells to confer protection. This was consistent with data showing that a TcR⁺ 'NK cell' clone also protected well against MCMV (53, 61). This cloned NK-like cell line is presumably of a similar lineage to the T-LAK cells. Similar experiments were done with the NK cell-resistant virus, LCMV, and neither the LAK cells nor the above mentioned NK-like, T-cell clone protected suckling mice from LCMV (53, 60) (Fig. 4.4).

LAK cells responded chemotactically to IFN and to exudates from virus-infected tissue (62), suggesting that they might home to sites of virus infection *in vivo* after implantation. This indeed was the case, as i.v.-injected ^{125}IUDR-labelled LAK cells did migrate preferentially into the peritoneal cavity of mice inoculated i.p. with MCMV, LCMV, and other viruses (64) (Fig. 4.5). However, although the accumulation of LAK cells in the virus-infected peritoneal cavity was 3–7 times that of control levels, the actual number of cells accumulating there was small, representing a low percentage (0.1–0.5 per cent) of the LAK cell inoculum. Analysis of the distribution of LAK cells in other organs led to the surprising observation that the virus-infected mice had significantly fewer LAK cell-associated CPM in other organs. Further analyses, mostly in the LCMV system, showed that LAK cells were rejected from virus-infected mice by an NK cell-dependent mechanism (63). Peak rejection occurred at the peak of the NK cell response during the course of infection, and the rejection was

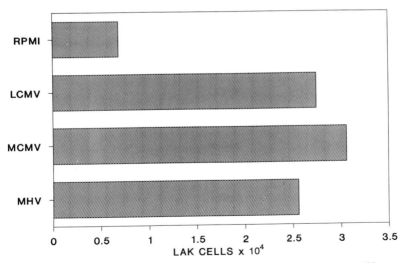

Fig. 4.5 Accumulation of LAK cells in virus-infected peritoneal cavity. ^{125}IUDR-radiolabelled lymphokine activated killer (LAK) cells (10^7) were inoculated i.v. into mice infected 3 days previously with viruses or control medium. Numbers represent the CPM-equivalent number of LAK cells migrated into the peritoneal cavity 2 h after inoculation. (This figure is based on results presented in (62).)

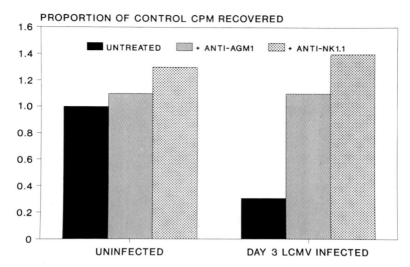

Fig. 4.6 Rejection of implanted lymphokine activated killer (LAK) cells by NK cells in virus-infected mice. [125]IUDR-radiolabelled LAK cells were injected i.v. into control mice or mice 3 days after i.p. LCMV inoculation. Some groups of mice were depleted of NK cell activity by treatments with antiserum to asialo-GM$_1$ or antibody to NK1.1. Data are expressed as the proportion of untreated controls. CPM recovered in the lung 4 h after injection of LAK cells. (This figure is based on data (63).)

abrogated by treatment of mice *in vivo* with antibodies to asialo-GM$_1$ or NK1.1 (Fig. 4.6). Thus, when inoculated into virus-infected mice, LAK cells do migrate to points of infection, but along the way their numbers are severely depleted by the activated NK cell response. This phenomenon could potentially provide a serious obstacle for LAK cell therapy of virus infections, and, indeed, effective LAK cell therapy for viral infections has yet to be reported.

5   Mechanisms for the selective lysis of virus-infected cells by NK cells

Theoretically, NK cells could control viral infections by a variety of mechanisms, including: (1) direct cell-mediated lysis of the virus-infected target cell, (2) a sublethal 'prelytic halt' of viral replication (an antiviral phenomenon proposed for CTL) (64), (3) the secretion of cytokines such as IFN and TNFα, which inhibit virus replication, and (4) the secretion of cytokines which influence other mechanisms of natural immunity, such as those mediated by macrophages. It is possible that one or a number of these mechanisms could play a role in the defence against a particular infectious agent, but most emphasis in this area has focused on the direct cytotoxic effects of NK cells.

Some evidence supports the concept that NK cells regulate viral infection *in vivo* by a cytolytic mechanism. Mice with the beige mutation have a defect in NK cell lytic activity but not NK cell number, and they have more difficulty in controlling the NK-sensitive MCMV than do their normal counterparts (46, 47, 49). However, most work suggesting a lytic mechanism comes from studies *in vitro* which have frequently indicated that NK cells can lyse virus-infected targets.

Evidence for the selective lysis of virus-infected cells will be presented below with the analysis of the various virus groups, but several fundamental concepts can be first outlined. The first is that the virus-infected cell may not have increased sensitivity to lysis, but it nevertheless may be killed somewhat selectively due to localized activation of NK cells. This activation could be mediated by the secretion of IFN and other NK cell-activating cytokines (65, 66), or it may be mediated by the triggering of NK cells by viral-encoded glycoproteins (9–13). In several infections (reviewed below) there appears to be both IFN-independent and IFN-dependent stages which contribute to the lysis of the virus-infected target cell. Virus-infected targets frequently compete well in cold target competition assays (67, 68, 69), which tend to measure triggering of effector cells and secretion of their cytolytic contents rather than binding of effectors to targets (69). Some virus infections, such as with Sendai, mumps, and vesicular stomatitis (VSV) viruses, enhance the binding of NK cells to targets (11, 68, 71), although this enhanced binding does not necessarily correlate with increased killing (68). Inhibitors of RNA and protein synthesis render cells more sensitive to NK cells by undefined mechanisms (72), and viral infections frequently result in similar inhibitions in cellular macromolecule synthesis. Little is known about the NK target structure, but some associations have been made between the expression of activation antigens, such as the transferrin receptor, which is induced by DNA virus, and increased sensitivity to NK cell mediated lysis (73–75). Inverse correlations have been noted between cell surface sialic acid and sensitivity to lysis (76, 77). This is of note because many viruses code for neuraminidase proteins which alter cellular sialation. Much recent interest has focused on the apparent inverse correlation between the expression of MHC Class I molecules and sensitivity to NK cells (78, 79). Reduced expression of Class I molecules is a frequent occurrence during viral infection, by either intracellular complexing of Class I molecules, selective inhibition of MHC gene transcription, or more global inhibitions in cellular protein synthesis (80).

5.1   Protection of target cells from NK cell killing by interferons

One phenomenon which may be of great importance *in vivo* is the ability of IFN α, β, and γ to protect target cells from NK cell-mediated lysis (7, 68, 81). The paradox of the NK cell system is that the same substance which acts as the major activating agent for NK cells also protects target cells from NK cell-mediated lysis. The mechanism of protection is somewhat unclear and may involve several mechanisms, such as down-regulation of activation antigens like the transferrin receptor (73) and up-regulation of cell surface sialation (77), but the most convincing correlation to date has been its influence on MHC antigen expression. Mutant cell lines deficient in Class I MHC expression are not protected against NK cells by IFN, even though IFN can establish antiviral and growth inhibitory effects in these same cells (78, 79). IFN protection is not an *in vitro* artefact, since thymocytes and tumour cells isolated from virus-infected or IFN-treated mice are resistant to NK cells (82, 83), and IFN-treated tumour cells are rejected *in vivo* by NK cells at a reduced rate (81). During a viral infection *in vivo*, IFN and possibly other cytokines, such as TNFα and IL-4, condition cells to become more resistant to NK cells as they up-regulate their MHC antigens and become more sensitive to MHC-restricted T-cells (82, 83). This makes sense, as it would be undesirable for the activated non-specific NK cells to attack normal cells in the body, and it would be desirable to up-regulate MHC antigens, which are expressed at low levels in most tissues, such that the cells could be adequately surveyed by T-cells. Many viral infections, by virtue of their non-specific alterations of cellular RNA and protein synthesis, render cells resistant to the effects of IFN, which require *de novo* RNA and protein synthesis (7). Thus, cells that first get infected with a virus would not be protected by IFN, whereas neighbouring uninfected cells would, thereby enabling the activated NK cells to selectively lyse virus-infected but not uninfected cells *in vivo* (84). Consistent with this hypothesis is that the NK-sensitive MCMV renders cells resistant to IFN protection, whereas this is not true for the less cytopathic NK-resistant LCMV (85).

5.2   Antibody-dependent cell-mediated cytotoxicity (ADCC)

One mechanism by which NK cells can selectively lyse virus-infected cells is by the ADCC mechanism late in infection when appropriate antiviral antibodies have been produced. NK cell-mediated ADCC has been demonstrated in a variety of viral systems (Table 4.1), but the significance of ADCC in the regulation of viral infections for the most part remains to be determined. Correlations have been made between the ability of mice

Table 4.1   Human antibody-dependent cell-mediated cytotoxicity to virus-infected cells

Herpesvirus	*Paramyxovirus*
Herpes simplex virus (HSV) (86)	Mumps virus (86)
Varicella zoster virus (VZV) (87)	Measles virus (86)
Epstein–Barr virus (EBV) (88)	Respiratory sincitial virus (RSV) (86)
Orthomyxovirus	*Poxvirus*
Influenza virus (86)	Vaccinia virus (55)
Flavivirus	*Retrovirus*
Dengue virus (89)	Human immunodeficiency virus-1 (HIV-1) (90–93)
	Human immunodeficiency virus-2 (HIV-2) (94)
	Human T-lymphotropic virus-1 (HTLV-1) (90)

Numbers in brackets indicate references.

to generate ADCC antibody and their susceptibility to herpes simplex virus-1 (95), and between the abilities of immunodeficient humans to mediate ADCC and to control viral infections (96). Furthermore, adoptive transfer of human mononuclear cells, human IFN, and appropriate antibodies mediating ADCC into suckling mice protected them against HSV-1 infection (97, 98). The relative roles of NK cell-mediated ADCC, monocyte- or granulocyte-mediated ADCC, and antibody and complement-mediated lysis late in viral infections have unfortunately still not been established in most systems.

6   Overview of the antiviral effects of NK cells in diverse viral systems

The potential roles for NK cells have been evaluated in a number of viral systems using *in vitro* cytotoxicity assays, clinical correlates, and animal model systems. Below is a summary of the information available in the better-studied systems.

6.1   Herpesviruses

6.1.1   Alpha herpesviruses

The alpha herpesviruses include HSV-1 and 2, varicella zoster virus (VZV), and a variety of animal viruses, including bovine herpes virus

(BHV-1). Interest in the herpesviruses has stemmed from the observations that NK cells lyse HSV-1-infected target cells (65, 68, 91, 99–101), that NK cells inhibit the progression of HSV-1 infection in infected target cells *in vitro* (102, 103), and that individuals with a propensity to get severe HSV-1 infections tend to have an NK cell response which is poor at lysing HSV-1-infected targets. These individuals include neonates, genetically immuno-deficient patients with Wiskott–Aldrich syndrome, and patients with im-munocompromising diseases such as Hodgkin's disease and AIDS (103–105). Of interest is a 1989 report of a patient bearing a complete absence of $CD16^+$ NK cell number, NK cell activity, and IL-2-dependent LAK activity, but having an otherwise normal immune system (106). Over a period of several years this patient presented first with a severe life-threatening VZV infection which required treatment with acyclovir, then with a severe CMV infection, which required intensive care for maintenance, and later with a severe HSV lesion, which also was treated with acyclovir. In addition, many of the stress-related factors which have been associated with reactivation of herpes labialis or herpes genitalis are associated with depressions in the NK cell response.

Human, mouse, and bovine NK cells lyse target cells infected with herpes viruses at high efficiency (65, 68, 99–101). Most of these studies have involved overnight cytotoxicity assays which provide time for effector cells to become activated, and effector cells removed from these assays have been shown to be activated, as judged by their ability to lyse uninfected targets. There is little evidence that HSV-1-infected target cells are innately more sensitive to lysis by NK cells, and two reports indicate that cells infected with HSV can actually become more resistant to NK cells when experiments are designed to bypass activation of the effector cells in the assays (68, 107, 108). Substantial levels of IFN are generated during the cytotoxicity assays, implying that natural killing against HSV-1-infected targets is simply a function of virus-induced IFN-dependent activation of NK cells (65, 66, 109, 110). However, the kinetics of lysis do not correlate closely with IFN production, and lysis is not blocked by including antibody to IFN or even actinomycin D (which globally blocks mRNA synthesis) into the cytoxicity assays (109, 110). $TNF\alpha$ and $TNF\beta$ are also produced in these assays, but antibodies to these cytokines also did not block cyto-toxicity, nor did HSV-1-infected fibroblasts demonstrate enhanced sensi-tivity to $TNF\alpha$ (111).

Nevertheless, the activation of NK cells seems to be important for anti-HSV-1 cytotoxicity, and more recent work in both the human HSV-1 and CMV systems has documented the importance of an accessory cell that is likely to send an activation signal to the NK cells in these assays. This accessory cell does not express antigens characteristic for NK, T- or B-cells, but does express high levels of Class II MHC antigens and has a number of properties in common with dendritic cells (112, 113). This accessory cell is responsible for much of the IFNα produced in HSV-1 cytotoxicity assays (113), and it is possible, but has not yet been demonstrated, that the accessory cell delivers IFN to the NK cell by some form of cell-to-cell contact that would protect the IFN from neutralization by antibody. The accessory cell and the NK cell are not only phenotypically distinct but under different regulatory control, the result being that different individuals have different relative levels of killing against NK cell-sensitive lymphomas and HSV-1-infected fibroblasts (113).

Several studies have examined the relative roles of viral gene products in these cytotoxicity reactions. Treatment of HSV-1- or BHV-infected cells with phosphonoformic or phosphonoacetic acid to block herpes DNA replication and late gene expression did not inhibit cytotoxicity, indicating that early, rather than late viral genes were involved (75, 114). Similar results were demonstrated with HCMV-infected targets (115). The expression of the transferrin receptor, which is a candidate target structure for NK cells, is induced by HSV and HCMV early genes (73–75). Antibody to this receptor inhibited the killing of HSV-1-infected targets, but purified transferrin did not inhibit the lysis of HCMV-infected targets. Some data indicate that cell surface expression of HSV-1 proteins is important for the 'natural killing' mechanism(s). Cells infected with an HSV-1 mutant in the late gene-encoded glycoprotein gC were lysed poorly, and monoclonal antibodies to gB (an early protein) and to gC blocked killing (116, 117). Some HSV-1 mutants selected for non-reactivity with these antibodies resisted 'natural killing'. This work argues for late as well as early gene-encoded proteins being of significance, but is in need of confirmation. Recent work with BHV has used the strategy of transfecting cells with vectors expressing viral gene products. Transfected cells expressing the BHV-I and -IV early genes (equivalent to HSV-1 gB and gD) displayed enhanced sensitivity to lysis, whereas cells expressing the BHV-III gene (equivalent to HSV-1 gC) were not more sensitive to lysis, again arguing for the importance of early genes (118). In these studies the roles of accessory cells were not addressed, so it is not clear whether the viral gene products were influencing accessory cell or killer cell function.

Whether NK cells play a significant role in HSV infections of the mouse has been controversial. Early work showed some genetic correlations between high NK cell activity and resistance to HSV-1 (119) and that depletion of NK cells with ^{89}strontium (120) or antiserum to asialo-GM$_1$ (121) enhanced the severity of infection. However, further analyses indicated that genetic resistance of mice to HSV-1 was primarily due to the early IFN response (that is, IFN induced during the first round of viral replication) (122–124), that NK cell-deficient beige mice were not very susceptible to HSV-1 (125), and that antiserum to asialo-GM$_1$ did not render mice more sensitive to HSV-1 except under conditions of very high doses, at which levels it also inhibited the IFN response (121, 125, 126). The genetic resistance to HSV-2 in mice does not correlate well with NK cell levels (127), and beige mice or mice depleted of NK cells are not more susceptible to HSV-2 (128). More recently, however, an adoptive transfer system has been developed to show that under certain conditions NK cells can play a role in murine herpes infections (129). Transfer of normal spleen cells into cyclophosphamide-treated mice conferred a protection against HSV-1 infection which was depleted by treating the transferred cells with complement and antiserum to asialo-GM_1 or monoclonal antibody NK 1.1. Also, 10-fold-fewer poly I:C-activated spleen cells were able to confer protection (129). A possible explanation for these above findings could be that both IFN and NK cells can play important roles in natural resistance to HSV-1, and that under normal conditions the degree of IFN-induced resistance independent of NK cells is so high that depletion of NK cells does not render a more severe infection. However, in the cyclophosphamide-treated animal, in which there might be an impairment of IFN production, including 'early IFN', NK cells would take on a greater role.

6.1.2   Beta herpesviruses

The beta herpesviruses include the cytomegaloviruses of man and animals. The substantial evidence that NK cells are responsible for natural immunity to MCMV is summarized above in Sections 3.1 and 4 of this review. HCMV-infected targets are readily lysed by human NK cells (130), and peripheral blood lymphocytes (PBL) bind better to HCMV-infected targets than to uninfected cells, as determined by direct analysis of effector cell/target cell conjugates (115). HCMV-infected targets become sensitive to killing before there is considerable cell-surface expression of viral glycoproteins (74, 115). Sensitivity to lysis correlated with the HCMV-induced expression of the cell transferrin receptor, but soluble transferrin did not block lysis (74). A recent finding, discussed above with the alpha herpesviruses, has shown that killing of HCMV-infected target cells by NK cells requires the interaction of $CD16^+NK$ cells with an $HLA-DR^+$ cell (112, 131). This accessory cell is reported to secrete IFNα which in turn activates the NK cell to mediate the cytolytic effect. This conclusion is supported by the observation that cytotoxicity here (in contrast to the HSV-1 system) was blocked by antibodies to IFNα (112). In short-term cytotoxicity assays, however, IFN does not seem to play a major role in the cytotoxicity mediated by NK cells against HCMV-infected targets (115).

6.1.3   Gamma herpesviruses

The gamma herpesviruses include the lymphotropic herpesviruses, whose most prominent member is Epstein–Barr virus (EBV). Lymphoblastoid cell lines stimulated into productive EBV infection by superinfection or by phorbol esters were killed to high levels by PBL from both seronegative and seropositive individuals (67). The killing of these cells appeared to be, at least in part, independent of IFN, as no IFN was detected in the culture supernatants. Cold target competition assays indicated that NK cells bound to, or were triggered by, virus-producing cell lines more efficiently than by non-producing cell lines (67). Low density, Percoll-separated, CD16$^+$ LGLs inhibited the *in vitro* proliferation of EBV-tansformed B-cells obtained from seropositive donors (132, 133), but it is not clear whether this was the result of cytotoxicity alone or whether other mechanisms, such as the production of growth inhibitory cytokines like IFN, were also involved. Chediak–Higashi syndrome (CHS), the human equivalent of the beige mouse, is a rare disease associated with very low NK activity (134). CHS patients frequently present with severe chronic infections with EBV (135). Severe EBV infections are also frequent among patients with X-linked lymphoproliferative syndrome who also have low levels of NK cell activity (136). However, diminished NK cell activity in these cases may be a consequence, rather than a factor predisposing to infection.

6.2   Paramyxoviruses

Mumps (10, 11), measles (9, 137), and Sendai (13) viruses have been extensively studied for their ability to induce and augment NK cell-mediated cytotoxicity, termed virus-dependent cell-mediated cytotoxicity (VDCC) (11). In these reactions, viral glycoproteins directly activate the NK effector cell. Purified mumps virus haemagglutinin (HN) but not fusion (F) protein efficiently induce VDCC. The specificity of this phenomenon was shown using monoclonal antibodies against mumps HN, and three serologically defined sites on the HN protein, distinct from those involved in hemagglutination, haemolysis, neuraminidase activity, and infectivity of the virus, were found to play a role in the activation of the NK cells (11). The phenomenon of VDCC thus seemed to require direct recognition of viral glycoproteins by the NK cell. Mumps' virions and purified glyco-proteins also augmented ADCC by a similar pathway. Both CD3$^-$ NK cells and CD3$^+$ 'non-MHC-restricted' T-cells mediated VDCC. This heterogeneity of effector cells was revealed using a single cell cytotoxicity assay which allowed for the analysis of the phenotype of the lymphocytes in effector cell/target cell conjugates rather than in bulk cell popula-tions (138). It was concluded that different effector cells were selected depending on the target cell against which cytotoxicity was measured (139).

Purified glycoproteins from the measles virus have been shown to stimu-late MHC-unrestricted cytotoxicity by human PBL (9), and measles virus-infected targets are sensitive to NK cell-mediated lysis *in vitro* (137). The sensitivity of measles virus-infected targets to NK cells was subdivided into two stages, an early (4 h) stage what was blocked by antibodies to measles but not to IFN, and a late (overnight) stage in which there was substantially enhanced killing that is blocked by antibodies to IFN. This suggested that a direct triggering of NK cells by viral glycoproteins expressed on the infected target cell may account for the enhanced killing in a 4 h assay and that the eventual production of IFN, and subsequent further activation of the NK cell population, may have accounted for the killing in more prolonged assays (137).

Sendai virus-infected cells have been reported to be killed to a greater degree than uninfected cells, but this seems to be dependent on the cell type and duration of infection, as in certain conditions infected cells are actually rendered more resistant to lysis (68, 140, 141). Sendai virus has the ability to bind and fuse membranes of many cell types, and NK cells efficiently conjugate with Sendai virus-infected cells (68, 140, 141). The latter compete efficiently in cold target competition assays and inhibit cytotoxicity of NK cells against NK-sensitive target cells. The enhanced binding of NK cells to Sendai virus-infected cells may account for the enhanced sensitivity of these cells to NK cell-mediated killing, but enhanced binding can occur in this system without enhanced killing (68).

6.3   Orthomyxoviruses

The influenza virus system provided the first demonstration of NK cell activation during a defined human virus infection (142). Influenza virus-infected target cells are killed by NK cells in long-term cytotoxicity assays (longer than 12 h) by a mechanism that involves the activation of NK cells by IFN (65, 66). In addition, however, the influenza neuraminidase (NA) possesses an intrinsic potential to stimulate NK-like cytotoxicity (12). The stimulatory site on the molecule appears to be different from the enzymatic site of NA (143). The non-structural (NS1) glycoprotein and haemagglutinin (HA) of influenza derived by cloning on *Escherichia coli* also enhance cytotoxicity mediated by Percoll-enriched LGL (144). In these experiments, effector cells were incubated for 18 h in the presence of purified proteins, after which cytotoxicity was measured against several target cells. The elevated cytotoxicity displayed by the cells was the result of activation of the effector cells by IFNα (and IFNγ) produced during the co-cultivation of the cells with the influenza virus proteins (144).

Influenza virus induces a rather typical NK cell response in mice, inducing high levels of NK cell activity in the lung (25), but it is unclear how much of the NK cell activation is due to IFN and how much to direct effects of the viral glycoproteins. Injections of purified influenza virus HA and NA activated mouse NK cells *in vivo* (145). Studies using antiserum to asialo-GM_1 to deplete NK cells in mice and hamsters infected intrathecally (i.t.) with influenza virus indicated much greater synthesis of virus and disease pathology (146). Controlled experiments done in the same laboratory showed that depletion of NK cells by intraperitoneal (i.p.) inoculation of anti-asialo-GM$_1$ resulted in more severe infections in mice infected by the i.t., but not by the intranasal (i.n.) route. However, if anti-asialo-GM$_1$ was also administered i.n., the infection was more severe when virus was inoculated by the i.n. route (147). This experiment suggests that NK cells in the upper respiratory system may be inaccessible to anti-NK cell antibodies administered by traditional routes.

6.4 Rhabdoviruses

The major work on the control of rhabdovirus infections by NK cells has been with vesicular stomatitis virus (VSV). Although there is little evidence that NK cells directly control VSV infection in mice, VSV has nevertheless received a great deal of study due to its effects on the growth of persistently infected tumour cells. Certain tumour cell lines with relatively non-cytopathic persistent infections with VSV, measles, and other viruses fail to form tumours in nude mice (69, 148, 149). Rejection of the tumours from nude mice was likely to be NK cell-mediated, as it was blocked by irradiation and antibodies to IFN (which block NK cell activation) or to asialo-GM$_1$. Different variants of VSV were isolated and used to initiate persistent infections. In general, there was a correlation between the ability of the variant to induce tumour rejection *in vivo* and the sensitivity of cells infected with that variant to NK cells *in vitro* (150–152). The antitumour effect seemed to be directed against the virus-infected target and was not a generalized antitumour response due to systemic NK cell activation, because other tumour cells implanted at different sites in the same mice were not rejected. Sequence comparisons between the NK-resistant and NK-sensitive variants revealed many amino acid differences in the G protein and several in the M and NS proteins (151).

The mechanism by which VSV renders cells more sensitive to NK cell-mediated lysis is not well understood, but VSV-infected targets competed well in cold target competition assays, suggesting that there may be enhanced binding or triggering of NK cells (69), and some studies examining conjugate formation suggested that there was enhanced binding (68, 71). Cells infected with VSV ts mutants in G, M, or L genes were not sensitive to lysis at the non-permissive temperature (71). However, antibodies to VSV proteins did not block lysis, and transfected cells expressing the VSV G gene were not rendered more susceptible to lysis (71). The data, therefore, support the interpretation that a productive VSV infection was needed for sensitivity to lysis, instead of the G, M, or L proteins being targets for the NK cells.

6.5   Arenaviruses

The evidence that the arenavirus LCMV is resistant to NK cells has been summarized in Sections 3.2 and 4. However, the arenavirus Pichinde (PV), which induces an LCMV-like NK cell and CTL response in mice, behaves quite differently to LCMV with regards to its sensitivity to NK cells. Antibodies to asialo-GM_1 or NK1.1 enhance PV replication in adult mice (51). Adoptive transfer of splenocytes from adult mice into suckling mice provides some protection to PV, and this is partially abrogated by antibody to asialo-GM1 (153). Subclones of PV, strain AN3739 were found to have, varying sensitivities to NK cells *in vivo*. The most NK-sensitive variant was passaged into mice treated with polyI:C to activate NK cells, and from these mice were isolated variants of PV that were resistant to NK cells but not in sensitivity to IFN (M. Vargas-Cortes and R. M. Welsh, unpublished observations). This indicates that NK-resistant variants can arise from NK-sensitive ones, and analysis of these and other such variants will help clarify the question of what factors determine NK-sensitivity of viruses *in vivo*.

6.6   Picornaviruses

Picornaviruses are highly cytopathic viruses which fail to express cell surface glycoproteins but nevertheless may alter the cell surface, as shown by altered lectin-binding properties (154). Coxsackie, encephalomyocarditis (EMCV), and Theiler's murine encephalitis (TMEV) viruses all induce T-cell-mediated pathology late in infection, associated with various degrees of insulitis, myocarditis, and demyelination. Treatment of mice with antiserum to asialo-GM_1 enhances the synthesis of Coxsackie virus and ultimately leads to more severe virus-induced cytopathic effect and T-cell-dependent immunopathology occurring as a consequence of the greater number of cells which become infected with virus (155). Adoptive transfer of cells 3 days post-infection into adult mice protected the recipients, and asialo-GM_1^+ cells were required for protection (156). A similar response was observed in mice infected with the diabetogenic (D) strain of EMCV. Injection of normally resistant C57BL/6 mice with antiserum to asialo-GM_1 resulted in high titres of the virus and a high incidence of diabetes after EMCV-D infection (157). Of interest in the EMCV system is a sex-linked resistance of female mice to EMCV which correlates with the ability of NK cells in these mice to release IFNγ (32, 33). Depletion of NK cells in C57BL/6 mice using either antiserum to asialo-GM_1 or the monoclonal antibody, NK1.1, led to an order of magnitude more severe inflammation of the brain gray matter (neuronal) 6 days after infection of mice with TMEV, and TMEV-infected beige mice also had slightly more severe lesions than infected control mice (158). Viral titres were not examined early during the TMEV infection to assess the role of NK cells in regulating the synthesis of virus *in vivo*. Studies *in vitro* indicated that Coxsackie virus-infected cells, perhaps because of the virus-induced alterations in cellular lectin-binding properties, were very sensitive to mouse NK cells (154, 156). Additionally, human NK cells have also been reported to lyse targets infected with the picornavirus hepatitis A (159). From the above data it appears clear that NK cells provide natural resistance to picorna-viruses, which are 'naked' viruses that do not display significant levels of viral proteins on the cell membrane. It is tempting to speculate that re-duced NK cell activity caused by stress might render a subject more sensi-tive to rhinoviruses, which are picornaviruses that cause the common cold.

6.7 Coronaviruses

Some reports indicate that infection of mice with the coronavirus mouse hepatitis virus (MHV) is enhanced by treatment with antiserum to asialo GM_1 (48, 160). However, we were unable recently to repeat our own experi-ments in this regard with a new stock of virus which had been passaged *in vivo* (51). Here, mice treated with anti-asialo-GM_1 or anti-NK1.1 had only slight enhancements of MHV synthesis. Quite possibly *in vivo*-passage selected for resistance to NK cells, as it did in controlled experiments for PV. Adoptive transfer studies in suckling mice with another strain of MHV indicated that resistance was partially conferred by a bone marrow cell which matured at the third week of life, consistent with an NK cell phenotype (161).

6.8 Flaviviruses and togaviruses

The NK cell-resistant Raji cell line when infected with the flavivirus dengue virus was killed by human PBL in long-term cytotoxicity assays, without the apparent participation of IFN (89). Analysis of effector cell phenotypes in this system revealed that some of the killing was mediated by prototypical NK cells and part by non-MHC-restricted CTL. The ratio of Dengue virus-associated $CD3^-$ to $CD3^+$ killing depended on the target cell type (89), similar to findings in the paramyxovirus VDCC system. Little is known about the sensitivity of togaviruses to NK cells, but the replication of the alpha togavirus Sindbis appears normal in NK cell-deficient beige mice (162), even though NK cells are prominent cells in the leucocyte infiltrate of the cerebrospinal fluid early in infection (27).

6.9 Poxviruses

Mice depleted of NK cells with antiserum to asialo-GM_1 synthesized higher amounts of vaccinia virus (VV) and ectromelia virus (46, 163, 164), and beige mice were more sensitive than normal mice to infection with ectromelia virus (163). Injection of athymic nude or normal mice with a VV recombinant encoding the IL-2 gene stimulated a potent NK cell response, and this virus was cleared much more rapidly than the control virus (165, 166). As IL-2 can provide chemotactic, activation, and proliferation signals for NK cells, it is possible that NK cells controlled this infection in the absence of a T-cell response.

6.10 Adenoviruses and papovaviruses

Adenovirus-induced enhancement of susceptibility to NK cells has been studied in several species. Inverse correlations were noted between the sensitivity of adenovirus-transformed cells to NK cells *in vitro* and their abilities to grow as tumours in rats (167). In the hamster model, the highly oncogenic adenovirus A12-infected cells were relatively resistant to NK cells, whereas the non-oncogenic adenovirus A2-infected cells were sensitive to NK cells (168). In rat, hamster, and mouse NK cell assays of cells infected with a variety of mutants of the non-oncogenic adenoviruses A2 and A5, the ability to induce sensitivity to NK cells mapped to the E1A gene (169, 170). Mutations in the E1B, E3, or E4 genes did not affect susceptibility to NK cells. The E1A region encodes a cell growth-promoting oncogene-like molecule, but transfection of cells in this system with the *myc* and the *ras* oncogenes did not mediate a similar effect (171). Strong correlations were noted between the expression of the E1A gene product and the susceptibility of mouse cells to lysis by TNFα, and the lysis of these targets by NK cells was partially inhibited by antibody to TNFα (172). Another relevant factor in this system may rest with the observation that class I MHC antigens can be down-regulated by adenovirus infection (80). In one study, human cells infected with adenovirus strains 2 or 12 were both rendered more susceptible to NK cells in a pattern that correlated with decreased Class I MHC antigen expression (173). In addition to enhanced killing there was also an enhanced binding of the NK cells to the virus-infected targets (173). Studies on the NK-sensitivity of papovavirus-infected cells are less extensive, but mouse fibroblasts transfected with the gene encoding the simian virus 40 T antigen also develop increased sensitivity to NK cells (170).

6.11 Retroviruses

Cells transformed by murine retroviruses are often quite sensitive to NK cells, the most notable example being the Moloney murine leukaemia virus (M-MuLV) infected YAC-1 cells, which are the prototypic NK cell targets in the mouse (174). Early workers in the field had speculated that

MuLV-encoded proteins may be the targets for NK cells, but this was ruled out by additional experimentation [for review, see (2)]. There have been few experiments designed to clearly determine whether NK cells provide natural resistance to murine retroviral infections. Resistance to the Friend virus complex is mediated by a ^{89}Sr-sensitive bone marrow cell (175) and is low in NK cell-deficient beige mice (176). Conditions of split-dose irradiation which deplete NK cell activity stimulate a retrovirus-dependent thymic leukaemia (177).

In human studies, both unstimulated and IL-2-stimulated NK cells have been shown to be capable of lysing freshly isolated T-cells and T-cell lines infected with human T-cell leukaemia virus (HTLV-I) or HIV (178). These studies also showed that LGLs protect T-cells from infection by HTLV-I, most likely by a mechanism that involves the participation of IFN. By contrast, LGLs were unable to protect T-cells against infection by HIV (178). Some, but not all, CD4$^+$ T-cell lines infected with HIV-1 were killed to high levels by PBL from healthy donors (179, 180). There appeared to be no correlation between the expression of HIV antigens on the target cells and the cytotoxicity mediated by NK cells against them (179). The killing of some HIV-infected cells by NK cells required the presence of an IFNα-producing HLA-Dr$^+$ cell similar to that observed in the HSV-1 and HCMV systems (179). Other cell lines were killed well in the absence of this accessory cell (179). Together, these reports give support to the hypothesis that NK cells may in fact be of some value in controlling human retroviral infections.

7   Viral infections of NK cells and virus-induced down-regulation of NK cell activity

Infection of leucocytes or of purified NK cells with HCMV, measles virus, or HIV *in vitro*, causes a suppression in NK cell activity. Clinical isolates of HCMV were shown to profoundly reduce NK cell activity without affecting ADCC (181, 182). This suppression of NK activity required live virus and a relatively long (more than 72 h) period of cultivation of the cells with the virus (181). MCMV-induced down-regulation of NK cells was partially restored by IFN, but not by IL-2 (181). Suppression did not occur when monocytes were depleted from the cell population, and the addition of infected monocytes to uninfected lymphocyte cultures induced suppression (181). These results suggested an important role of monocytes and monocyte-derived factors in the induction of suppression of NK cells in HCMV infection. Exposure of human PBL to the measles virus resulted in an infection of NK cells which did not alter their viability or ADCC capacity, but did reduce their ability to directly lyse NK-sensitive targets (183). It is not clear whether such a measles virus-induced suppression of

NK cell activity was a direct consequence of the infection of NK cells or due to an indirect effect on the modulating functions of accessory cells for NK cells, as observed with MCMV. Human retroviruses as well as retroviruses from diverse species are commonly immunosuppressive and oncogenic in their hosts (184, 185). Some of the mechanisms by which HIV may suppress NK cell activity are discussed below in Section 8.

8   Analysis of the NK cell and ADCC response in AIDS

HIV is the etiological agent for AIDS, which is characterized by the development of Kaposi's sarcoma and severe opportunistic infections by, amongst others, the NK cell-sensitive herpes viruses (186). NK cell activity is reduced in AIDS and patients with AIDS-related complex (ARC) but is essentially normal in asymptomatic, seropositive individuals (184, 187). Extensive studies have been aimed at identifying the nature of the defect of NK cells in AIDS patients. Most reports have shown that NK cells in AIDS patients are phenotypically and numerically normal (188–191), although one report (192), described a selective depletion in the number of $CD16^+/CD56^+$ NK cells which also expressed the CD8 antigen. NK cells from AIDS patients were found to be defective in their ability to reorganize the microtubule-organizing center (191). When bound to targets the NK cells from AIDS patients showed normal polarization of actin but alpha and beta tubulin failed to undergo polarization. Actin filaments appear to be involved in target cell binding, whereas tubulin is involved in microtubule reorganization (191). In one series of studies NK cells from AIDS or ARC patients bound well to NK-sensitive target cells, but failed to secrete natural killer cell cytotoxic factors (NKCF) upon stimulation with the appropriate target (193). This defect was partially restored by exposure to IL-2. NKCF was released from NK cells of AIDS and ARC patients upon stimulation with phorbol esters and calcium ionophore (193). In contrast to the NK cell response, ADCC activity in such patients was found to be within the normal range. Two-target conjugation assays, in which an NK and an ADCC target were used, demonstrated that the AIDS effector cells lysed only the bound ADCC target, whereas the normal effector cells lysed both the bound NK and ADCC targets. Moreover, AIDS NK cells stimulated with ADCC targets released NKCF (194). These data suggest that NK cells in AIDS patients are triggered normally for ADCC but are not triggered for NK cell activity, and that the lytic machinery of the cells is not significantly impaired.

HIV-specific antibodies capable of inducing normal levels of HIV-specific ADCC were found in sera from HIV-seropositive individuals (90–94). By using target cells that expressed different HIV proteins after infection with constructs of vaccinia virus as vectors, it was shown that ADCC was directed against envelope glycoproteins, but not against gag proteins (93). One study indicated that both NK activity and HIV-specific ADCC declined with disease progression. Here, the impairment of ADCC was due to both a defect in the cells that mediate ADCC and to a defect in the ability to generate anti-HIV antibodies functional in ADCC (195).

The exact manner by which NK cells are affected in AIDS patients is not known, and much of the work in this area must be viewed with caution, as contaminating mycoplasma could account for some of the HIV-related effects *in vitro*. It is unlikely that the reduction in NK cell activity is the result of infection of the NK cell by HIV, although there is a report that IL-2-cultured CD16$^+$ NK cells in the presence of HIV lost lytic activity and expressed HIV antigens (196). However, there is no direct evidence that CD4$^-$ LGLs are infected by HIV *in vivo*, but rather that an indirect pathogenic mechanism is most likely responsible for NK cell disfunction in AIDS. Of interest is the observation that some synthetic peptides homologous to HIV-1 envelope glycoproteins can inhibit NK cell mediated killing *in vitro*. Peptides corresponding to regions located in the transmembrane region of gp 41 of HIV inhibited NK cell activity against K562 cells. The NK cells, however, were able to bind target cells in a normal fashion, indicating that inhibition occurred at a post-binding step in the cytolytic reaction (197). This led to the suggestion that the envelope glycoproteins might modulate NK cell function, in the absence of HIV infection of the NK cell (184). Another interesting possibility derives from the finding that the vasoactive intestinal peptide (VIP) shares an amino acid homology with CD4 at the binding site for gp 120. VIP inhibited the binding of purified gp 120 from HIV to neuronal cells, and the gp 120 induced killing of neuronal cells (198). VIP also inhibited the chemotaxis induced by gp 120 on CD4$^+$ T-cells (198). NK cells were shown to have receptors for VIP (200), and VIP inhibited the NK cell activity of normal PBL (201) and caused an increase in intracellular levels of cAMP (202). Increased cAMP levels correlate with a diminished activity of NK cells from normal individuals (203). It is thus plausible that interactions between VIP receptors on the NK cells and gp 120 may be responsible for the defect of the NK cells in AIDS. Despite intensive study, it still remains to be settled which, if any, of these mechanisms is responsible for the alteration of the NK cells in AIDS and ARC patients, and how important this depression in NK cell activity is for the control of the HIV infection and the adventitious agents.

9 Summary

It can be concluded that viral infections profoundly stimulate the activation and proliferation of NK cells and that NK cells provide natural resistance to many, but not all, viruses. The isolation of NK 'escape' viral variants suggests that the interaction between NK cells and viruses may be a changing, dynamic process, particularly during a persistent infection. The means by which virus-infected cells can be lysed by NK cells appear multifold, varying with the virus and the target cell. *In situ* activation of the NK cell may be mediated by either certain viral glycoproteins and/or virus-induced IFN, sometimes produced by a dendritic-like accessory cell. Some viruses may render cells innately more sensitive to NK cells by down-regulating target cell Class I MHC expression, or by a variety of other mechanisms which might enhance NK cell binding or triggering, or inhibit cellular repair or resistance functions. The selective protection of un-infected, but not virus-infected cells *in vivo* by IFN may enable activated NK cells to selectively direct their cytotoxicity against the virus-infected targets. How important NK cells are to the natural immunity of humans to viruses remains to be clearly resolved, but severe infections are common-place in NK cell-deficient individuals, and studies in murine models attest to the significance of NK cells at early stages of infection. It could perhaps be concluded that, for many infections, NK cells determine whether one gets sick and the T-cells determine whether one gets better.

References

1　Fitzgerald-Bocarsly, P. *et al.* (1988). A definition of natural killer cells. *Immunol. Today*, **9**, 292.

2　Welsh, R. M. (1981). Natural killer cells in virus infections. *Curr. Top. Microbiol. Immunol.*, **92**, 83–106.

3　Welsh, R. M. (1986). Regulation of virus infections by cytotoxic leukocytes. *Clin. Immunol. Newslett.*, **7**, 171–5.

4　Welsh, R. M. (1978). Cytotoxic cells induced during lymphocytic choriomeningitis virus infection of mice. I. Characterization of natural killer cell induction. *J. Exp. Med.*, **148**, 163–81.

5　Welsh, R. M. (1984). Natural killer cells and interferon. *CRC Crit. Rev. Immunol.*, **5**, 55–93.

6　Gidlund, M., Orn, A., Wigzell, H. Senik, A., and Gresser, I. (1978). Enhanced NK activity in mice injected with interferon and interferon inducers. *Nature*, **273**, 759–61.

7　Trinchieri, G. and Santoli, D. (1978). Antiviral activity induced by culturing lymphocytes with tumor derived or virus-transformed cells. Enhancement of natural killer activity by interferon and antagonistic inhibition of susceptibility of target cells to lysis. *J. Exp. Med.*, **147**, 1314–33.

8　Kuribayashi, K., Gillis, S., Kern, D. E., and Henney, C. S. (1981).

Murine NK cell cultures: effects of interleukin-2 and interferon on cell growth and cytotoxic reactivity. *J. Immunol.* **126**, 2321–7.

9 Casali, P., Sissons, J. G. P., Buchmeier, M. J., and Oldstone, M. B. A. (1981). *In vitro* generation of human cytotoxic lymphocytes by virus. Viral glycoproteins induce nonspecific cell-mediated cytotoxicity without release of interferon. *J. Exp. Med.*, **154**, 840–55.

10 Harfast, B., Orvell, C., Alsheikhly, A., Andersson, T., Perlmann, P., and Norrby, E. (1980). The role of viral glycoproteins in mumps virus-dependent lymphocyte-mediated cytotoxicity *in vitro*. *Scand. J. Immunol.*, **11**, 391–400.

11 Alsheikhly, A-R., Orvell, C., Andersson, T., and Perlmann, P. (1985). The role of serologically defined epitopes on mumps virus HN-glycoproteins in the induction of virus-dependent cell mediated cytotoxicity (VDCC) *in vitro*. Analysis with monoclonal antibodies. *Scand. J. Immunol.*, **22**, 529–38.

12 Arora, D. J. S., House, M., Justewicz, D. M., and Mandeville, R. (1984). *In vitro* enhancement of human natural cell-mediated cytotxicity by purified influenza virus glycoproteins. *J. Virol.*, **52**, 839–45.

13 Alsheikhly, A., Orvell, C., Harfast, B., Andersson, T., Perlmann, P., and Norrby, E. (1983). Sendai virus induced cell mediated cytotoxicity *in vitro*. The role of viral glycoproteins in cell mediated cytotoxicity. *Scand. J. Immunol.*, **17**, 129–38.

14 Welsh, R. M., Brubaker, J. O., Vargas-Cortes, M., and O'Donnell, C. L. Natural killer cell response to virus infections in mice with severe combined immunodeficiency (SCID). The stimulation of NK cells and the NK cell-dependent control of virus infections occur independently of T and B cell function. *J. Exp. Med.*, **173**, 1053–63.

15 Miller, S. C. (1982). Production and renewal of murine natural killer cells in the spleen and bone marrow. *J. Immunol.*, **129**, 2282–6.

16 Biron, C. A., Turgiss, L. R., and Welsh, R. M. (1983). Increase in NK cell number and turnover rate during acute viral infection. J. Immunol., **131**, 1539–45.

17 Kiessling, R., Eriksson, E., Hallenbeck, L. A., and Welsh, R. M. (1980). A comparative analysis of the cell surface properties of activated versus endogenous mouse natural killer cells. *J. Immunol.*, **125**, 1551–7.

18 Herberman, R. B., Bartram, S., Haskill, J. A., Nunn, M., Holden, H. T., and West, W. H. (1977). Fc receptors on mouse effector cells mediating natural cytotoxicity against tumor cells. *J. Immunol.*, **119**, 322–6.

19 Biron, C. A. and Welsh, R. M. (1982). Blastogenesis of natural killer cells during viral infection *in vivo*. *J. Immunol.*, **129**, 2788–95.

20 Biron, C. A., Sonnenfeld, G., and Welsh, R. M. (1984). Interferon

induces natural killer cell blastogenesis *in vivo. J. Leuc. Biol.*, **35**, 31–7.

21 Gresser, I., Bandu, M.-T., and Brouty-Boyte, D. (1974). Interferon and cell division. IX. Interferon-resistant L1210 cells: characteristics and origin. *J. Nat. Cancer Inst.*, **52**, 553–9.

22 Natuk, R. J. and Welsh, R. M. (1987). Accumulation and chemotaxis of large granular lymphocytes at sites of virus replication. *J. Immunol.*, **138**, 877–83.

23 McIntyre, K. W. and Welsh, R. M. (1986). Accumulation of natural killer and cytotoxic T large granular lymphocytes in the livers of virus-infected mice. *J. Exp. Med.*, **164**, 1667–81.

24 Welsh, R. M. (1986). Regulation of virus infections by natural killer cells. *Nat. Immun. Cell Growth Regul.*, **5**, 169–99.

25 Stein-Streilein, J., Bennett, M., Mann D., and Kumar, V. (1983). Natural killer cells in mouse lung: surface phenotype target preference, and response to local influenza virus infection. *J. Immunol.*, **131**, 2699–2704.

26 Doherty, P. C. and Korngold, R. (1983). Characteristics of poxvirus-induced meningitis: virus-specific and non-specific cytotoxic effectors in the inflammatory exudate. *Scand. J. Immunol*, **18**, 1–7.

27 Griffin, D. E. and Hess, J. L. (1986). Cells with natural killer activity in the cerebrospinal fluid of normal mice and athymic nude mice with acute Sindbis virus encephalitis. *J. Immunol.*, **136**, 1841–5.

28 McIntyre, K. W., Natuk, R. J., Biron, C. A., Kase, K., Greenberger, J., and Welsh, R. M. (1988). Blastogenesis of large granular lymphocytes in non-lymphoid organs. *J. Leuc. Biol.*, **43**, 492–501.

29 Wiltrout, R. H., *et al.* (1989). Augmentation of mouse liver-associated natural killer cell activity by biological response modifiers occurs largely via rapid recruitment of large granular lymphoctyes from the bone marrow. *J. Immunol.*, **143**, 372–8.

30 Natuk, R. J. and Welsh, R. M. (1987). Chemotactic effect of human recombinant interleukin-2 on mouse activated large granular lymphocytes. *J. Immunol.*, **139**, 2737–43.

31 Goldfield, A. E. and Maniatis, T. (1989). Coordinate viral induction of tumor necrosis factor α and interferon β in human B cells and monocytes. *Proc. Nat. Acad. Sci. USA*, **86**, 1490–4.

32 McFarland, H. I. and Bigley, N. J. (1989). Sex-dependent, early cytokine production by NK-like spleen cells following infection with the D variant of encephalomyocardidtis virus (EMCV-D). *Viral Immunol.*, **2**, 205–214.

33 McFarland, H. I. and Bigley, N. J. (1990). AGM1+ spleen cells contain gamma interferon (IFN-τ) gene transcripts in the early, sex-dependent production of IFN-after picornavirus infection. *J. Virol.*, **64**, 4407–13.

34 Kobayashi, M., *et al.* (1989). Identification and purification of natural killer cell stimulatory factor (NKSF), a cytokine with multiple biological effects on human lympocytes. *J. Exp. Med.*, **170**, 827–46.

35 Suzuki, R., Handa, K., Itoh, K., and Kumagai, K. (1983). Natural killer (NK) cells as a responder to interleukin-2 (IL-2). I. Proliferative response and establishment of cloned cells. *J. Immunol.*, **130**, 981–7.

36 Biron, C. A. Young, H. A., and Kasaian, M. T. (1990). Interleukin 2-induced proliferation of murine natural killer cells *in vivo*. *J. Exp. Med.*, **171**, 173–88.

37 Migliorati, G., Cannarile, L., Herberman, R. B., Bartocci, A., Stanley, E. R. and Riccardi, C. (1987). Role of interleukin 2 (IL-2) and hemopoietin (H1) in the generation of mouse natural killer (NK) cells from primitive bone marrow precursors. *J. Immunol.*, **138**, 3618–25.

38 van den Brink, M. R. M., Boggs, S. S., Herberman, R. B., and Hiserodt, J. C. (1990). The generation of natural killer (NK) cells from NK precursor cells in rat long-term bone marrow cultures. *J. Exp. Med.*, **172**, 303–14.

39 Ishida, Y., *et al.* (1989). Expansion of natural killer cells but not T cells in human interleukin 2/interleukin 2 receptor (Tac) transgenic mice. *J. Exp. Med.*, **170**, 1103–15.

40 Kasaian, M. T. and Biron, C. A. (1989). The activation of IL-2 transcription in L3T4$^+$ and LYT-2$^+$ lymphocytes during virus infection *in vivo*. *J. Immunol.*, **142**, 1287–92.

41 Kasaian, M. T. and Biron, C. A. (1990). Cyclosporin A inhibition of interleukin 2 gene expression, but not natural killer cell proliferation, after interferon induction *in vivo J. Exp. Med.*, **171**, 745–62.

42 Staczek, J. (1990). Animal cytomegaloviruses. *Microbiol. Rev.*, **54**, 247–65.

43 Boos, J. and Wheelock, E. F. (1971). Correlation of survival from MCMV infection with spleen cell responsiveness to conconavalin A. *Proc. Soc. Exp. Biol. Med.*, **149**, 443–6.

44 Kiessling, R., Klein, E., Pross, H., and Wigzell, H. (1975). 'Natural' killer cells in the mouse. II Cytotoxic cells with specificity for mouse Moloney leukemia cells. Characteristic of the killer cell. *Eur. J. Immunol.*, **5**, 117–21.

45 Bancroft, G. J., Shellam, G. R., and Chalmer, J. E. (1981). Genetic influences on the augmentation of natural killer (NK) cells during murine cytomegalovirus infection. Correlation with patterns of resistance. *J. Immunol.*, **126**, 988–94.

46 Roder, J. and Duwe, A. (1979). The beige mutation in the mouse selectivity impairs natural killer cell function. *Nature*, **278**, 451–3.

47 Shellam, G. R., Allan, J. E., Papadimitriou, J. M., and Bancroft, G. J. (1981). Increased susceptibility to cytomegalovirus infection in beige mutant mice. *Proc. Nat. Acad. Sci. U.S.A.*, **78**, 5104–8.

48 Bukowski, J. F., Woda, B. A., Habu, S., Okumura, K., and Welsh, R. M. (1983). Natural killer cell depletion enhances virus synthesis and virus-induced hepatitis *in vivo*. *J. Immunol.*, **131**, 1531–8.

49 Bukowski, J. F. Woda, B. A. and Welsh, R. M. (1984). Pathogenesis of murine cytomegalovirus infection in natural killer cell-depleted mice. *J. Virol.*, **52**, 119–28.

50 Ebihara, K. and Minamishima, Y. (1984). Protective effect of biological response modifiers on murine cytomegalovirus infection. *J. Virol.*, **51**, 117–22.

51 Welsh, R. M., Dundon, P. L., Eynon, E. E., Brubaker, J. O., Koo, G. C., and O'Donnell, C. L. (1990). Demonstration of the antiviral role of natural killer cells *in vivo* with a natural killer cell-specific monoclonal antibody. *Nat. Immun. Cell Growth Regul.*, **9**, 112–20.

52 Shanley, J. D. (1990). *In vivo* administration of monoclonal antibody to the NK1.1 antigen of natural killer cells; effect on acute murine cytomegalovirus infection. *J. Med. Virol.*, **30**, 58–60.

53 Bukowski, J. F., Warner, J. F., Dennert, G., and Welsh, R. M. (1985). Adoptive transfer studies demonstrating the antiviral effects of NK cells *in vivo*. *J. Exp. Med.*, **161**, 40–52.

54 Pape, G., Troye, M., Axelsson, B., and Perlmann, P. (1979). Simultaneous occurrence of immunoglobulin-dependent and immunoglobulin-independent mechanisms in natural cytotoxicity of human lymphocytes. *J. Immunol.*, **122**, 2251–60.

55 Perrin, L. H., Zinkernagel, R. M., and Oldstone, M. B. A. (1977). Immune response in humans after vaccination with vaccinia virus: generation of a virus-specific cytotoxic activity by human peripheral lymphocytes. *J. Exp. med.*, **146**, 949–69.

56 Welsh, R. M. and Kiessling, R. W. (1980). Natural killer cell response to lymphocytic choriomeningitis virus in beige mice. *Scand. J. Immunol.*, **11**, 363–7.

57 Welsh, R. M., Biron, C. A., Bukowski, J. F., McIntyre, K. W., and Yang, H. (1984). Role of natural killer cells in virus infections of mice. *Surv. Synth. Path. Res.*, **3**, 409–31.

58 Bukowski, J. F., Biron, C. A., and Welsh, R. M. (1983). Elevated natural killer cell-mediated cytotoxicity, plasma interferon, and tumor cell rejection in mice persistently infected with lymphocytic choriomeningitis virus. *J. Immunol.*, **131**, 991–6.

59 Grimm, E. A., Mazumder, A., Zhang, H. Z., and Rosenberg, S. A. (1982). Lymphokine-activated killer cell phenomenon. Lysis of natural killer resistant fresh solid tumor cells by interleukin 2-activated autologous human peripheral blood lymphocytes. *J. Exp. Med.*, **155**, 1823–41.

60 Bukowski, J. F., Yang, H., and Welsh, R. M. (1988). The antiviral

effect of lymphokine activated killer cells. I. Characterization of the effector cells mediating prophylaxis. *J. Virol.*, **62**, 3642–8.

61 Yanagi, Y., *et al.* (1985). Gene rearrangement in cells with natural killer activity and expression of the β-chain of the T-cell antigen receptor. *Nature*, **314**, 631–3.

62 Natuk, R. J., Bukowski, J. F., Brubaker, J., and Welsh, R. M. (1989). Antiviral effects of lymphokine-activated killer (LAK) cells. 2. Chemotaxis and homing to sites of infection. *J. Virol.*, **63**, 4969–71.

63 Brubaker, J. P., Chong, K. T., and Welsh, R. M. (1991). Lymphokine-activated killer (LAK) cells are rejected *in vivo* by activated natural killer cells. *J. Immunol.*, **147**, 1439–44.

64 Martz, E. and Howell, D. (1989). CTL: virus control cells first and cytolytic cells second? *Immunol. Today*, **10**, 79–86.

65 Santoli, D., Trinchieri, G., and Lief, F. S. (1978). Cell-mediated cytotoxicity against virus-infected target cells in humans. I. Characterization of the effector lymphocyte. *J. Immunol.*, **121**, 526–31.

66 Santoli, D., Trinchieri, G., and Koprowski, H. (1978). Cell-mediated cytotoxicity against virus-infected target cells in humans. II. Interferon induction and activation of natural killer cells. *J. Immunol.*, **121**, 532–8.

67 Blazar, B., Patarroyo, M., Klein, E., and Klein, G. (1980). Increased sensitivity of human lymphoid lines to natural killer cells after induction of the Epstein–Barr viral cycle by superinfection or sodium butyrate. *J. Exp. Med.*, **151**, 614–27.

68 Welsh, R. M. and Hallenbeck, L. A. (1980). Effect of virus infections on target cell susceptibility to natural killer cell-mediated lysis. *J. Immunol.*, **124**, 2491–7.

69 Minato, N., Bloom, B. R., Jones, C., Holland, J., and Reid, L. M. (1979). Mechanism of rejection of virus in persistently infected tumor cells by athymic nude mice. *J. Exp. Med.*, **149**, 1117–33.

70 Perussia, B. and Trinchieri, G. (1981). Inactivation of natural killer cell cytotoxic activity after interaction with target cells. *J. Immunol.* **126**, 754–8.

71 Moller, J. R., (1985). Natural killer cell recognition of target cells expressing different antigens of vesicular stomatitis virus. *Proc. Natl. Acad. Sci. U.S.A.*, **82**, 2456–9.

72 Kunkel, L. A. and Welsh, R. M. (1981). Metabolic inhibitors render "resistant" target cells sensitive to natural killer cell-mediated lysis. *Int. J. Cancer*, **27**, 73–9.

73 Vodinelich, L., Sutherland, R., Schneider, C., Newman, R., and Greaves, M. F. (1983). Receptor for transferrin may be a "target" structure for natural killer cells. *Proc. Natl. Acad. Sci. U.S.A.*, **80**, 835–9.

74 Borysiewicz, L. K., Graham, S., and Sissons, J. G. (1986). Human

natural killer cell lysis of virus-infected cells. Relationship to expression of the transferrin receptor. *Eur. J. Immunol.*, **16**, 405–11.

75 Lopez-Guerrero, J. A., Alarcon, B., and Fresno, M. (1988). Mechanism of recognition of herpes simplex virus type 1-infected cells by natural killer cells. *J. Gen. Virol.*, **69**, 2859–68.

76 Yogeeswaran, G., Gronberg, A., Hansson, M., Dalianis, T., Kiessling, R., and Welsh, R. M. (1981). Correlation of glycosphingolipids and sialic acid in YAC-1 lymphoma variants with their sensitivity to natural killer cell mediated lysis. *Int. J. Cancer*, **28**, 517–26.

77 Yogeeswaran, G., Fujinami, R., Kiessling, R., and Welsh, R. M. (1982). Interferon-induced alterations in cellular sialic acid and glycoconjugates. Correlation with susceptibility to activated natural killer cells. *Virology*, **121**, 363–71.

78 Ljunggren, H.-G. and Karre, K. (1990). In search of the 'missing self': MHC molecules and NK cell recognition. *Immunol. Today*, **11**, 237–44.

79 Piontek, G. E., *et al.* (1985). YAC-1 MHC class I variants reveal an association between decreased NK sensitivity and increased H-2 expression after interferon treatment or *in vivo* passage. *J. Immunol.*, **135**, 4281–8.

80 Schrier, P. I., Bernards, R., Vaessen, R. T. M. J., Houweling, A., and van der Eb, A. J. (1983). Expression of class 1 major histocompatibility antigens switched off by highly oncogenic adenovirus 12 in transformed rat cells. *Nature (London)*, **305**, 771–5.

81 Welsh, R. M., Karre, K., Hansson, M., Kunkel, L. A., and Kiessling, R. W. (1981). Interferon-mediated protection of normal and tumor target cells against lysis by mouse natural killer cells. *J. Immunol.*, **126**, 219–25.

82 Hansson, M., Kiessling, R., Andersson, B., and Welsh, R. M. (1980). Effect of interferon and interferon inducers on the NK sensitivity of normal mouse thymocytes. *J. Immunol.*, **125**, 2225–31.

83 Bukowski, J. F. and Welsh, R. M. (1986). Enhanced susceptibility to cytotoxic T lymphocytes of target cells isolated from virus-infected or interferon-treated mice. *J. Virol.*, **59**, 735–9.

84 Santoli, D., and Koprowski, H. 1979). Mechanisms of activation of human natural killer cells against tumor and virus-infected cells. *Immunol. Rev.*, **44**, 125–63.

85 Bukowski, J. F. and Welsh, R. M. (1985). Inability of interferon to protect virus-infected cells against lysis by natural killer (NK) cells correlates with NK cell-mediated antiviral effects *in vivo*. *J. Immunol.*, **135**, 3537–41.

86 Sissons, J. G. P. and Oldstone M. B. A. (1980). Antibody-mediated destruction of virus infected cells. **29**, 209–60.

87 Ito, M., Ihara, T., Grose, C., and Starr, S. (1985). Human leukocytes kill varicella zoster virus infected fibroblasts in the presence of murine

monoclonal antibodies to virus specific glycoproteins. *J. Virol.*, **54,** 98–103.

88 Pearson, G. R. and Orr, T. W. (1976). Antibody-dependent lymphocyte cytotoxicity against cells expressing Epstein–Barr virus antigen. *J. Nat. Cancer Inst.*, **56,** 485–8.

89 Kurane, I., Hebblewaite, D., Brandt, W. E., and Ennis, F. E. (1984). Lysis of Dengue virus-infected cells by natural cell-mediated cytotoxicity and antibody-dependent cell-mediated cytotoxicity. *J. Virol.*, **52,** 223–30.

90 Sinclair, A. L., *et al.* (1988). Antibody-dependent cell-mediated cytotoxicity: comparison between HTLV-I and HIV-1 assays. *AIDS*, **2,** 465–72.

91 Rook, A. H., Lane, H. C., Folks, T., McCoy, S., Alter, H., and Fauci, A.S. (1987). Sera from HTLV-III/LAV antibody-positive individuals mediate antibody-dependent cellular cytotoxicity against HTLV-III/LAV infected T cells. *J. Immunol.*, **138,** 1064–7.

92 Ljunggren, K., Bottinger, B., Biberfeld, G., Karlson, A., Fenjo, E. M., and Jondal, M. (1987). Antibody-dependent cell mediated cytotoxicity-inducing antibodies against human immunodeficiency virus. Presence at different clinical stages. *J. Immunol.*, **139,** 2263–7.

93 Koup, R. A., *et al.* (189). Antigenic specificity of antibody-dependent cell-mediated cytotoxicity against human immunodeficiency virus in antibody-positive sera. *J. Virol.*, **63,** 584–90.

94 Norley, S. G., Mikschy, U., Werner, A., Staszewski, S., Helm, E. B., and Kurth, R. (1990). Demonstration of cross-reactive antibodies able to elicit lysis of both HIV-1 and HIV-2 infected cells. *J. Immunol.*, **145,** 1700–5.

95 Kohl, S., Thomas, J. W., and Loo, L.-S. (1986). Defective production of anti-herpes simplex antibody by neonatal mice. Reconstitution with Ia$^+$ macrophages and T helper lymphocytes from nonimmune adult syngeneic mice. *J. Immunol.*, **136,** 3038–44.

96 Kohl, S., Loo, L. S., Schmalsteig, F. S., and Anderson, D. C. (1986). The genetic deficiency of leukocyte surface glycoprotein Mac-1, LFA-1, p150, 95 in humans is associated with defective antibody-dependent cellular cytotoxicity *in vitro* and defective protection against herpes simplex virus infection *in vivo*. *J. Immunol.*, **137,** 1688–94.

97 Kohl, S., Loo, L. S., and Greenberg, S. S. (1982). Protection of newborn mice from a lethal herpes simplex virus infection by human interferon, antibody, and leukocytes. *J. Immunol.*, **128,** 1107–11.

98 Kohl, S. and Loo, L. S. (1982). Protection of neonatal mice against herpes simplex virus infection: probable *in vivo* antibody-dependent cellular cytotoxicity. *J. Immunol.*, **129,** 370–6.

99 Ching, C. and Lopez, C. (1979). Natural killing of herpes virus type-1

infected target cells: normal human responses and influence of antiviral antibody. *Infect. Immunol.*, **26**, 49–56.

100 Cook, C. G., and Splitter, G. A. (1989). Characterization of bovine mononuclear cell populations with natural cytolytic activity against bovine herpesvirus 1-infected cells. *Cell. Immunol.*, **120**, 240–49.

101 Tilden, A. B., Cauda, R., Grossi, C. E., Balch, C. M., Lakeman, A. D., and Whitley, R. J. (1986). Demonstration of NK cell-mediated lysis of varicella-zoster virus (VZV)-infected cells: characterization of the effector cells. *J. Immunol.*, **136**, 4243–8.

102 Fitzgerald, P. A., Mendelsohn, M., and Lopez, C. (1985). Human natural killer cells limit replication of herpes simplex virus type 1 *in vitro. J. Immunol.*, **134**, 2666–72.

103 Leibson, P. J., Hunter-Lasszlo, M., Douvas, G. S., and Hayward, A. R. (1986). Impaired neonatal natural killer-cell activity to herpes simplex virus: decreased inhibition of viral replication and altered response to lymphokines. *J. Clin. Immunol.*, **6**, 216–24.

104 Lopez, C., *et al.* (1983). Correlation between low natural killing of fibroblasts infected with herpes simplex virus type 1 and susceptibility to herpesvirus infections. *J. Infect. Dis.*, **147**, 1030–35.

105 Lopez, C., Fitzgerald, P. A., and Siegal, F. P. (1983). Severe acquired immune deficiency syndrome in male homosexuals: diminished capacity to make interferon alpha *in vitro* associated with severe opportunistic infections. *J. Infect. Dis.*, **148**, 962–6.

106 Biron, C. A., Byron, K. S., and Sullivan, J. S. 1989). Severe herpes virus infections in an adolescent without natural killer cells. *N. Eng. J. Med.*, **320**, 1731–5.

107 Bishop, G. A., McCurry, L., Schwartz, S. A., and Glorioso, J. C. (1987). Activation of human natural killer cells by herpes simplex virus type 1-infected cells. *Intervirol.*, **28**, 78–88.

108 Munoz, A., Carrasco, L., and Fresno, M. (1983). Enhancement of susceptibility of HSV-1 infected cells to natural killer cells by interferon. *J. Immunol.*, **131**, 783–7.

109 Fitzgerald, P. A. von Wussov, P., and Lopez, C. (1982). Role of interferon in the natural kill of HSV-1 infected fibroblasts. *J. Immunol.*, **129**, 819–23.

110 Bishop, G. A., Glorioso, J. C., and Schwartz, S. A. (1983). Role of interferon in human natural killer activity against target cells infected with HSV-1. *J. Immunol.*, **131**, 1849–53.

111 Oh, S. H., Trinchieri, G., Bandyopadhyay, S., and Starr, S. E. (1990). Natural killer cell-mediated lysis of herpes simplex virus-infected fibroblasts: inability to detect soluble factors that contribute to lysis. *Cell. Immunol.*, **127**, 221–9.

112 Bandyopadhyay, S., Perussia, B., Trinchieri, G., Miller, D. S., and Starr, S. (1986). Requirement for HLA-DR$^+$ accessory cells in natural

killing of cytomegalovirus-infected fibroblasts. *J. Exp. Med.*, **164**, 180–95.

113 Fitzgerald-Bocarsly, P., Feldman, M., Mendelsohn, M., Curl, S., and Lopez, C. (1988). Human mononuclear cells which produce interferon-alpha during NK(HSV-FS) assays are HLA-DR positive cells distinct from cytolytic natural killer effectors. *J. Leuc. Biol.*, **43**, 323–34.

114 Cook, C. G., Letchworth, J., and Splitter, G. A. (1989). Bovine naturally cytotoxic cell activation against bovine herpes virus type 1-infected cells does not require late viral glycoproteins. *Immunol.*, **66**, 565–9.

115 Borysiewicz, L. K., Rodgers, B., Morris, S., Graham, S., and Sissons, J. G. P. (1985). Lysis of human cytomegalovirus infected fibroblasts by natural killer cells: demonstration of an interferon-independent component requiring expression of early viral proteins and characterization of effector cells. *J. Immunol.*, **134**, 2695–2701.

116 Bishop, G. A. Glorioso, J. C., and Schwartz, S. A. (1983). Relationship between expression of herpes simplex virus glycoproteins and susceptibility of target cells to human natural killer cell activity. *J. Exp. Med.*, **157**, 1544–61.

117 Bishop, G. A., Marlin, S. D., Schwartz, S. A., and Glorioso, J. C. (1984). Human natural killer cell recognition of herpes simplex virus type 1. Glycoproteins: specificity analysis with the use of monoclonal antibodies and antigenic variants. *J. Immunol.*, **133**, 2206–14.

118 Palmer, L. D., Leary, T. P., Wilson, D. M., and Splitter, G. A. (1990). Bovine natural killer cell responses against cell lines expressing recombinant herpesvirus type 1 glycoproteins. *J. Immunol.*, **145**, 1009–14.

119 Lopez, C. (1981). Resistance to herpes simplex virus type-1. In *Natural resistance to tumors and viruses* (ed. O. Haller), pp. 15–24. Springer, New York.

120 Lopez, C., Ryshke, R., and Bennett, M. 1980). Marrow dependent cells depleted by ^{89}Sr mediate genetic resistance to herpes simplex virus type 1 infections in mice. *Infect. Immunol.*, **28**, 1028–32.

121 Habu, S., Akamatsu, K., Tamaoki, N., and Okumura, K. (1984). *In vivo* significance of NK cells on resistance against virus (HSV-1) infections in mice. *J. Immunol.*, **133**, 2743–7.

122 Zawatzky, R., Gresser, I., Demayer, F., and Kirchner, H. (1982). The role of interferon in the resistance of C57B1/6 mice to various doses of herpes simplex virus type 1. *J. Infect. Dis.*, **146**, 405–10.

123 Engler, H., Zawatzky, R., Kirchner, H., and Armerding, D. (1982). Experimental infection of inbred mice with herpes simplex virus. IV. Comparison of interferon production and natural killer cell activation insusceptible and resistant adult mice. *Arch. Virol.*, **74**, 239–47.

124 Kirchner, H., Engler, H., Schroder, C. H., Zawatzky, R., and Storch,

E. (1983). Herpes simplex virus type 1-induced interferon production and activation of natural killer cells in mice. *J. Gen. Virol.*, **64**, 437–41.

125 Bukowski, J. F. and Welsh, R. M. (1986). The role of natural killer cells and interferon in resistance to acute infection of mice with herpes simplex virus type 1. *J. Immunol.*, **137**, 3481–85.

126 Chmielgrczyk, W., Engler, H., Ernst, R., Opitz, U., and Kirchner, H. (1985). Injection of anti-thy 1.2 serum breaks genetic resistance of mice against herpes simplex virus. *J. Gen. Virol.*, **66**, 1087–94.

127 Ammerding, D. and Rossiter, H. (1981). Induction of natural killer cells by herpes simplex virus type 2 in resistant and sensitive inbred mouse strains. *Immunobiol.*, **158**, 369–79.

128 Morahan, P. S., Coleman, P. H., Morse, S. S., and Volkman, A. (1982). Resistance to infections in mice with defects in the activities of mononuclear phagocytes and natural killer cells: effects of immuno-modulators in beige mice and ^{89}Sr-treated mice. *Infect. Immun.*, **37**, 1079–85.

129 Rager-Zisman, B., Quan, P.-C., Rosner, M., Moller, J. R., and Bloom, B. R. (1987). Role of NK cells in protection of mice against herpes simplex virus-1 infection. *J. Immunol.*, **138**, 884–8.

130 Diamond, R. D., Keller, R., Lee, G., and Finkel, D. (1977). Lysis of cytomegalovirus-infected human fibroblasts and transformed human cells by peripheral blood lymphoid cells from normal human donors. *Proc. Soc. Exp. Biol. Med.*, **154**, 259–63.

131 Bandyopadhyay, S., *et al.* (1989). Natural killer cell mediated lysis of target cells infected with cytomegalovirus or human immunodeficiency virus. In *Natural killer cells and host defense* (ed. E. W. Ades and C. Lopez), pp. 114–19. Karger, Basel.

132 Shope, T. C. and Kaplan, J. (1979). Inhibition of the *in vitro* outgrowth of Epstein–Barr-virus-infected lymphocytes by T$_G$ lymphocytes. *J. Immunol.*, **123**, 2150–5.

133 Masucci, M. G., Bejarano, M. T., Masucci, G., and Klein. E. (1983). Large granular lymphocytes inhibit the *in vitro* growth of autologous Epstein–Barr-virus-infected B cells. *Cell. Immunol.*, **76**, 311–21.

134 Haliotis, T., *et al.* Chediak–Higashi gene in humans. I. Impairment of natural killer function. *J. Exp. Med.*, **151**, 1039–48.

135 Merino, F., Henle, W., and Ramirez-Duque, P. (1986). Chronic active Epstein–Barr virus infection in patients with Chediak–Higashi syndrome. *J. Clin. Immunol.*, **6**, 299–305.

136 Sullivan, J. L., Byron, B., Brewster, F. E., and Purtilo, D. T. (1980). Deficient natural killer cell activity in the X-linked lymphoproliferative syndrome. *Science*, **210**, 543–5.

137 Casali, P. and Oldstone, M. B. A. (1982). Mechanisms of killing of measles virus-infected cells by human lymphocytes: Interferon-

associated and unassociated cell-mediated cytotoxicity. *Cell. Immunol.*, **70**, 330–44.

138 Vargas-Cortes, M., Hellstrom, U., and Perlmann, P. (1983). Surface markers of human natural killer cells as analyzed in a modified single cell cytotoxicity assay on poly-L-lysine coated cover slips. *J. Immunol. Methods*, **62**, 87–99.

139 Alsheikhly, A. R., Andersson, T., and Perlmann, P. (1985). Virus dependent cellular cytotoxicity *in vitro*. Mechanism of induction and effector cell characterization. *Scand. J. Immunol.*, **21**, 329–35.

140 Anderson, M. J. 1978). Innate cytotoxicity of CBA mouse spleen cells to Sendai virus-infected L cells. *Infect. Immun.*, **20**, 608–12.

141 Weston, P. A., Jensen, P. J. Levy, N. L., and Koren, H. S. (1981). Spontaneous cytotoxicity against virus infected cells: Relationship to NK against uninfected cell line and to ADCC. *J. Immunol.*, **126**, 1220–4.

142 Ennis, F. A., *et al.* (1981). Interferon induction and increased natural killer cell activity in influenza infections in man. *Lancet*, **ii**, 891–93.

143 Arora, D. J. S., and Justewicz, D. M. (1988). Human influenza viral neuraminidases augment cell mediated cytotoxicity *in vitro*. Nat. Immun. Cell Growth Regulat., **7**, 87–94.

144 Rees, R. C., Dalton, B. J., Yonay, J. F., Hanna, N., and Poste, G. (1987). Augmentation of human natural killer cell activity by influenza virus antigens produced in *Escherichia coli. J. Biol. Resp. Modif.*, **6**, 69–87.

145 Arora, D. J. S. and Houde, M. (1988). Purified glycoproteins of influenza virus stimulate cell-mediated cytotoxicity *in vivo*. *Nat. Immun. Cell Growth Regul.*, **7**, 287–96.

146 Sten-Streilein, J. and Guffee, J. (1986). *In vivo* treatment of mice and hamsters with antibodies to asialo GM1 increases morbidity and mortality to pulmonary influenza infection. *J. Immunol.*, **136**, 1435–41.

147 Guffee, J., Fan, W. and Stein-Streilein, J. (1989). In *Natural killer cells and host defense* (ed. C. Lopez and E. W. Ades), pp. 124–30. Karger, Basel.

148 Reid, L. M., Jones, C. L. and Holland, J. (1979). Virus carrier state suppresses tumorigenicity of tumor cells in athymic (nude) mice. *J. Gen. Viro.*, **42**, 609–14.

149 Reid, L. M., Minato, N., Gresser, I., Holland, J., Kadish, A., and Bloom, B. R. (1981). Influence of anti-mouse interferon serum on the growth and metastasis of tumor cells persistently infected with virus and human prostatic tumors in athymic nude mice. *Proc. Natl. Acad. Sci. U.S.A..*, **78**, 1171–5.

150 Jones, C. L., Spindler, K. R., and Holland, J. J. (1980). Studies on tumorigenicity of cells persistently infected with visicular stomatitis virus in nude mice. *Virology*, **103**, 158–66.

151 Vandepol, S. B. and Holland, J. J. (1986). Evolution of vesicular stomatitis virus in athymic nude mice: mutations associated with natural killer cell selection. *J. Gen. Virol.*, **67**, 441–51.

152 Vandepol, S. B. and Holland, J. J. (1986). Tumorigenicity of persistently infected tumors in nude mice is a function of both the virus and the host cell type. *J. Virol.*, **58**, 914–20.

153 Welsh, R. M., Natuk, R. J., McIntyre, K. W., Yang, H., Biron, C. A., and Bukowski, J. F. (1988). Factors influencing the control of virus infections by natural killer cells. In *Functions of the natural immune system* (ed. C. W. Reynolds and R. H. Wiltrout), pp. 111–28. Plenum, New York.

154 Lutton, C. W. and Gauntt, C. J. (1986). Coxsackievirus B3 infection alters plasma membrane of neonatal skin fibroblasts. *J. Virol.*, **60**, 294–6.

155 Godeny, E. K. and Gauntt, C. J. (1986). Involvement of natural killer cells in coxsackievirus B3 viral-induced myocarditis. *J. Immunol.*, **137**, 1695–1702.

156 Godeny, E. K. and Gauntt, C. J. (1987). Murine natural killer cells limit coxsackievirus B3 replication. *J. Immunol.*, **139**, 913–18.

157 White, L. L. and Smith, R. A. (1990). D variant of encephalomyocarditis virus (EMCV-D)-induced diabetes following natural killer cell depletion in diabetes-resistant male C57BL/6J mice. *Viral Immunol.*, **3**, 67–76.

158 Paya, C. V., Patick, A. K., Leibson, P. J., and Rodriguez, M. (1989). Role of natural killer cells as immune effectors in encephalitis and demyelination induced by Theiler's virus. *J. Immunol.*, **143**, 95–102.

159 Kurane, I., Binn, L. N., Bancroft, W. H., and Ennis, F. E. (1985). Human lymphocyte response to hepatitis A virus-infected cells: Interferon production and lysis of infected cells. *J. Immunol.*, **135**, 2140–4.

160 Carman, P. S., Ernst, P. B., Rosenthal, K. L., Clark, D. A., Befus, A. D., and Bienenstock, J. (1986). Intraepithelial leukocytes contain a unique subpopulation of NK-like cytotoxic cells active in the defense of gut epithelium to enteric murine coronavirus. *J. Immunol.*, **136**, 1548–53.

161 Tardieu, M., Hery, C., and Dupuy, J. M. (1980). Neonatal susceptibility to MHV_3 infection in mice. II. Role of natural effector marrow cells in transfer of resistance. *J. Immunol.*, **124**, 428–23.

162 Hirsch, R. L. (1981). Natural killer cells appear to play no role in the recovery of mice from Sindbis virus infection. *Immunology*, **43**, 81–9.

163 Jacoby, R. O., Bhatt, P. N., and Brownstein, D. G. (1989). Evidence that NK cells and interferon are required for genetic resistance to lethal infection with ectromelia virus. *Arch. Virol.*, **108**, 49–58.

164 Bukowski, J. F., McIntyre, K., Yang, H., and Welsh, R. M. (1987). Natural killer cells are not required for interferon-mediated prophy-

laxis against vaccinia or murine cytomegalovirus infections. *J. Gen. Virol.*, **68**, 2219–22.

165 Ramshaw, I. A., Andrew, M. E., Phillips, S. M., Boyle, D. B., and Coupar, B. E. H. (1987). Recovery of immunodeficient mice from a vaccinia virus/IL-2 recombinant infection. *Nature*, **329**, 545–6.

166 Karupiah, G., *et al.* (1990). Elevated natural killer cell responses in mice infected with recombinant vaccinia virus encoding murine IL-2. *J. Immunol*, **144**, 290–8.

167 Raska, K. and Gillimore, P. H. (1982). An inverse relation of the oncogenic potential of adenovirus-transformed cells and their sensitivity to killing by syngeneic natural killer cells. *Virology*, **123**, 8–18.

168 Cook, J. L., and Lewis, Jr., A. M. (1984). Differential NK cell and macrophage killing of hamster cells infected with nononcogenic or oncogenic adenovirus. *Science*, **224**, 612–15.

169 Cook, J. L., May, D. L., Lewis, Jr., A. M., and Walker, T. A. (1987). Adenovirus E1A gene induction of susceptibility to lysis by natural killer cells and activated macrophages in rodent cells. *J. Virol.*, **61**, 3510–20.

170 Fresa, K. L., Karjalainen, H. E., and Tevethia, S. S. (1987). Sensitivity of simian virus 40-transformed C57BL/6 mouse embryo fibroblasts to lysis by murine natural killer cells. *J. Immunol.*, **138**, 1215–20.

171 Cook, J. L., May, D. L., Wilson, B. A., and Walker, T. A. (1989). Differential induction of cytolytic susceptibility by E1A, *myc*, and *ras* oncogenes in immortalized cells. *J. Virol.*, **63**, 3408–15.

172 Cook, J. L., *et al.* (1989). Role of tumor necrosis factor-alpha in E1A oncogene-induced susceptibility of neoplastic cells to lysis by natural killer cells and activated macrophages. *J. Immunol.*, **142**, 4527–34.

173 Dawson, J. R., Storkus, W. J., Patterson, E. B., and Cresswell, P. (1989). Adenovirus inversely modulates target cell class I MHC antigen expression and sensitivity to natural killing. In *Natural killer cells and host defense* (ed. E. W. Ades and C. Lopez, pp. 156–9. Karger, Basel.

174 Fenyo, E. M., Klein, E., Klein, G., and Swiech, K. (1968). Selection of an immunoresistant Moloney lymphoma subline with decreased concentration of tumor-specific surface antigens. *J. Nat. Cancer Inst.*, **40**, 69–89.

175 Kumar V. and Bennett, M. (1981). Genetic resistance to Friend virus-induced erythroleukemia and immunosuppression. In *Natural resistance to tumors and viruses* (ed. O. Haller), *Current topics in microbiology and immunology*, Vol. 92, pp. 65–82. Springer, New York.

176 Eckner, R. J., Bennett, M., Hettrick, K. L., and Seidler, C. (1987). Natural killer cell suppression of Friend virus-induced preleukemic hemopoietic stem cells. *J. Virol.*, **61**, 2631–8.

177 Parkinson, D. R., Brightman, R. P., and Waksal, S. D. (1981).

Altered natural killer cell biology in C57BL/6 mice after leukemo-genic split-dose irradiation. *J. Immunol.*, **126**, (4), 1460–4.

178 Ruscetti, F. W., *et al.* (1986). Analysis of effector mechanisms against HTLV-I and HTLV-III infected lymphoid cells. *J. Immunol.*, **136**, 3619–24.

179 Bandyopadhyay, S., Ziegner, U., Campbell, D. E., Miller, D. S., Hoxie, J. A., and Starr, S. E. (1990). Natural killer cell mediated lysis of T cell lines chronically infected with HIV-1. *Clin. Exp. Immunol.*, **79**, 430–5.

180 Rappociolo, G., Toso, J. F., Torpey, D. J., Gupta, P., and Rinaldo Jr., C. R. (1989). Association of alpha interferon production with natural killer cell lysis of U937 cells infected with human immuno-deficiency virus. *J. Clin. Microbiol.*, **27**, 41–8.

181 Schrier, R. D., Rice, G. P. A., and Oldstone, M. B. A. (1986). Suppression of natural killer cell activity and T cell proliferation by fresh isolates of human cytomegalovirus. *J. Infect. Dis.* **153**, 1084–91.

182 Schrier, R. D. and Oldstone, M. B. A. (1986). Recent clinical isolates of cytomegalovirus suppress human leukocyte antigen-restricted cyto-toxic T lymphocyte activity. *J. Virol.*, **59**, 127–31.

183 Casali, P., Rice, G. P. A., and Oldstone, M. B. A. (1984). Viruses disrupt functions of human lymphocytes: Effects of measles virus and influenza virus on lymphocyte mediated killing and antibody produc-tion. *J. Exp. Med.*, **159**, 1322–37.

184 Sirianni, M. C., Tagliaferri, F., and Aiuti, F. (1990). Pathogenesis of the natural killer cell deficiency in AIDS. *Immunol. Today*, **11**, 81–2.

185 Dent, P. B. (1972). Immunodepression by oncogenic viruses. *Prog. Med. Virol.*, **14**, 1–35.

186 Ragona, G. and Sirianni, M. C. (1987). Imbalance of the Epstein–Barr virus host relationship in AIDS-related complex patients. *Cancer Detect. Prevent.* **1**, (Suppl.) 549–52.

187 Ortona, L., Tamburrini, E., Tumbarello, M., Ventura, G., and Cauda, R. (1988). Natural killer (NK) activity in patients with HIV infection. *Bolletino Inst. Sieroterapico Milanese*, **67**, 135–41.

188 Mitchell, W. M., Forti, R. L., Vogler, L. B., Lawton, A. R., and Gregg, C. R. (1984). Spontaneous and interferon resistant natural killer cell anergy in AIDS. *AIDS Res.*, **1**, 221–9.

189 Rook, A. H., *et al.* (1983). Interleukin-2 enhances the depressed natural killer and cytomegalovirus specific cytotoxic activities of lym-phocytes from patients with the acquired immunodeficiency syndrome. *J. Clin. Invest.*, **72**, 398–403.

190 Fontana, L., Sirianni, M. C., DeSanctis, G., Carbonari, M., Ensoli, B., and Aiuti, F. (1986). Deficiency of natural killer cell activity, but not of natural killer binding, in patients with lymphadenopathy syn-

drome positive for antibodies to HTLV-III. *Immunobiol.*, **171**, 425–35.

191 Sirianni, M. C., Soddu, S., Malorni, W. Arancia, G., Aiuti, F., and Soddus, S. (1988). Mechanism of defective natural killer cell activity in patients with AIDS is associated with defective distribution of tubulin. *J. Immunol.*, **140**, 2565–8.

192 Viullier, F., Brauw, N. E., Montagnier, L., and Dighiero, G. (1988). Selective depletion of low density $CD8^+$, $CD16^+$ lymphocytes during HIV infection. *AIDS Res. Hum. Retro.*, **4**, 121–9.

193 Bonavida, B., Katz, J., and Gottlieb, M. (1986). Mechanism of defective NK cell activity in patients with acquired immunodeficiency syndrome (AIDS) and AIDS-related complex. I. Defective trigger on NK cells for NKCF production by target cells and partial restoration by IL-2. *J. Immunol.*, **137**, 1157–63.

194 Katz, J. D., Mitsugatsu, R., Gottlieb, M. S., Lebow, L. T., and Bonavida, B. (1987). Mechanisms of defective NK cell activity in patients with acquired immunodeficiency syndrome (AIDS) and AIDS-related complex. II. Normal antibody-dependent cellular cytotoxicity (ADCC) mediated by effector cells defective in natural killer (NK) cytotoxicity. *J. Immunol.*, **139**, 55–60.

195 Tyler, D. S., *et al.* (1990). Alterations in antibody-dependent cellular cytotoxicity during the course of HIV-1 infection. Humoral and cellular defects. *J. Immunol.*, **144**, 3375–84.

196 Robinson Jr., W. E., Mitchell, W. M., Chambers, W. H., Schuffman, S. S., Montefiori, C., and Oeltmann, T. N. (1988). Natural killer cell infection and inactivation *in vitro* by the human immunodeficiency virus. *Hum. Pathol.*, **19**, 535–40.

197 Cauda, R., Tumbarello, M., Ortona, L., Kanda, P., Kennedy, R. C., and Chanh, T. C. (1988). Inhibition of normal human natural killer cell activity by human immunodeficiency virus synthetic transmembrane peptides. *Cell. Immunol.*, 115, 57–65.

198 Brenneman, D. E., *et al.* (1988). Neuronal cell killing by the envelope protein of HIV and its prevention by vasoactive intestinal peptide. *Nature*, **335**, 639–42.

199 Sacerdote, P., Ruff, M. R., and Pert, C. B. (1988). VIP 1-12 is a ligand for the CD4/human immunodeficiency virus receptor. *Ann. N.Y. Acad. Sci.*, **527**, 574–8.

200 Ottaway, C. A. (1988). Vasoactive intestinal peptide as a modulator of lymphocyte and immune function. *Ann. NY Acad. Scie.*, **527**, 486–500.

201 Rola-Pleszczynsky, M., Bolduc, D., and St-Pierre, S. (1985). The effects of vasoactive intestinal peptide in human natural killer cell function. *J. Immunol.*, **135**, 2569–73.

202 Said, S. I. (1984). Vasoactive intestinal polypeptide (VIP) current status. *Peptides*, **5**, 143–50.

203 Ullberg, M., Jondal, M., Lanefelt, F., and Fredholm, B. B. (1983). Inhibition of human NK cell cytotoxicity by induction of cyclic AMP depends on impaired target cell recognition. *Scand. J. Immunol.*, **17**, 365–73.

5 Natural killer cells in transplantation

P. G. HOGAN

1   Introduction

The outcome of transplantation and the occurrence of complications, such as infection and graft-versus-host disease (GVHD), may be influenced by the non-specific cytotoxicity and regulatory functions of natural killer (NK) cells. The possible interactions after transplantation between NK cells, bone marrow or solid organ graft cells and alloantigen-specific immune cells are discussed here. Experimental models of hybrid resistance and clinical bone marrow transplant data have shown that NK cells may exert antigraft effects early after transplantation, but in addition may be protective against opportunistic infections and malignancy. However, the role of NK cells in both GVHD and in rejection of solid organ grafts remains unclear.

2   Properties of NK cells relevant to transplantation

2.1   Cytotoxic mechanisms

Transplantation of allogeneic cells into a recipient triggers a complex series of cellular and humoral interactions between the graft and the new host. Foremost among the mechanisms leading to allograft rejection is the specific recognition of graft major histocompatibility complex (MHC) and minor histocompatibility (mH) antigens by host T- and B-cells (1–4). Elimination of donor cells hours after transplantation (hyperacute rejection) or within 5 days (accelerated rejection) is mediated by complement-fixing antibodies to MHC antigens produced in recipients sensitized by previous trans-fusions, pregnancies or transplants. During the subsequent 100 days, T-cells are primarily responsible for episodes of acute rejection whereas the etiology of the late onset vascular obliterative lesions of chronic re-jection is uncertain (1). Alloantigens expressed in high density on dendritic or lymphoid cells in solid grafts may activate up to 15 per cent of recipient T-cells, with or without the participation of antigen presenting cells (1, 3, 4). Graft injury may follow, both directly by CD8[+] (cytotoxic) T-cells, and indirectly by CD4[+] (helper) T-cell-derived cytokines which mediate macrophage activation, antibody production and inflammatory cell influx into the graft (1, 4). In the case of bone marrow transplants, circulating recipient CD8[+] T-cells obtained at the time of rejection mediate both direct cytotoxicity and colony inhibitory activity specific for either MHC or mH antigens on donor haemopoietic stem cells (5).

Natural killer (NK) cells mediate a spectrum of cytotoxic and regulatory activities which may also contribute to the outcome of transplantation. The ability of NK cells to rapidly recognize and lyse some foreign cell types without prior sensitization is central to their role as a mechanism for the first line defence of the host (6). This function is well documented for cytomegalovirus (CMV) infections (see Chapter 4 of this volume), where

NK cells are the only effector elements capable of recognizing and lysing CMV-infected cells until sufficient numbers of specific cytotoxic T-cells are generated (7). There is evidence to indicate that NK cells may also rapidly lyse infused allogeneic lymphocytes (8). NK cells and their interferon (IFN)-inducible precursor cells are strategically located for such encounters primarily in blood (9), with small populations also present in bone marrow (10), spleen (11), and lung (12), but not in gut-associated lymphoid tissue (13). Indirect evidence suggests that blood NK cells may be able to access sites of transplantation such as the bone marrow and vascularized solid organ grafts. NK cells adhere to vascular endothelium and exhibit chemotaxis in response to casein and C5a but, as yet, there has been no demonstration of their ability to migrate through vascular endothelium (14–16). The leucocyte cellular adhesion molecules (leuCAMs a, b, and c), intercellular adhesion molecule type 1 (ICAM-1), and the CD2 and CD58 molecules assist NK cells to contact and lyse tumour cells and are most likely responsible for NK cell endothelial adherence (17, 18). Interleukin 2 (IL-2) and possibly other cytokines may increase adherence of NK cells to vascular endothelium by up-regulating the expression of these adhesion molecules (15, 18).

Haemopoietic stem cell populations are directly susceptible to lysis by NK cells and, when infused into peripheral blood during bone marrow transplantation, may be significantly depleted by circulating host NK cells which survive the transplant procedure (19, 20). Mature parenchymal and endothelial cells in solid organ allografts are not directly susceptible to NK activity, but these cells may become susceptible to lysis in the presence of graft-specific antibodies by the mechanism of antibody-dependent cellular cytotoxicity (ADCC). Peripheral blood ADCC is mediated largely by NK cells which utilize low affinity Fc receptors to recognize and lyse antibody-labelled targets (21).

2.2 Regulatory mechanisms

In addition to direct cytotoxicity, NK cells of graft or host origin may regulate specific immunological attack against a successfully implanted graft. As potent secretors of IFNα, IFNγ, IL-2, and tumour necrosis factor alpha (TNFα), alloantigen-stimulated NK cells may support both the afferent and effector arms of a graft-specific immune response (6, 22–25). IFNα, IFNγ and TNFα up-regulate the presentation of Class I and II MHC antigens on lymphohaemopoietic, epithelial, and endothelial cells (26, 27), whereas IFNγ and IL-2 may support the clonal expansion, maturation, and effector functions of graft-specific T-cells engaged in transplant rejection. NK cells may also exert non-specific suppression of T-cell responses to alloantigens, possibly due to direct lysis of antigen-presenting cells or secretion of suppressive cytokines (24, 25).

Addition of NK cells to B-cell cultures *in vitro* usually results in suppression of antibody synthesis through direct cytotoxicity or the secretion of inhibitory cytokines, such as IFNγ (28). However, NK cells also secrete a B-cell growth factor which may assist the recovery of antibody production after bone marrow transplantation (29). Endothelial cell activation, coagulation abnormalities, and inflammatory cell infiltration in rejecting graft tissue may be potentiated by NK cell-derived pro-inflammatory mediators which include IL-1, TNFα and plasminogen activator (23, 30, 31).

The final mechanism which has been postulated for NK cell participation in regulation of transplantation involves antigen presentation. There are several hypotheses describing how T-cells interact with foreign MHC or mH antigens, but most postulate involvement of an antigen-presenting cell, which displays alloantigen directly or after intracellular processing (3). NK cell subsets bearing Class II MHC antigens have vigorous stimulating activity comparable to monocytes in autologous or allogeneic mixed lymphocyte reactions (MLRs) (32). Donor NK cells carried in the graft may initiate immunological graft rejection by their antigen-presenting function, or alternatively, host NK cells surviving immunosuppression may serve as targets for GVHD.

2.3   Lymphokine-activated killer cells

NK cells had been recognized for over 10 years before lymphokine-activated killer (LAK) cells were classified as a separate entity, distinguished primarily by their ability to lyse NK-resistant tumour cells (33, 34). LAK cells are generated by exposure of NK or T-cell populations to proliferative doses of IL-2, resulting in LAK effector cell populations displaying either NK or T-cell phenotypes (or a mixture of the two) (21). Current interest in LAK cells focuses on their antineoplastic potential, but high levels of LAK activity demonstrated against some normal cell types may be relevant to transplantation (35, 36). In the allogeneic MLR, secretion of IL-2 by responder T-cells is sufficient to generate high levels of LAK activity from T-cell precursors, with maximum cytotoxicity against tumour targets peaking 7 days after first exposure to alloantigen (34, 37). Cytotoxic T-cells with LAK activity may also retain specific alloreactivity mediated through the T-cell receptor (TCR). It is possible, therefore, that appropriately stimulated T-cells may simultaneously mediate non-specific LAK activity and alloreactivity, perhaps against different components of the same target cell (38). Significant levels of IL-2 released spontaneously after transplantation, or as a consequence of anti-rejection therapy (39, 40), create the potential for LAK cells to be generated *in vivo*. Following bone marrow transplantation, peripheral blood lymphocytes (PBL) with spontaneous LAK activity exist *in vivo*, but their significance has not been explored in detail (41). There is limited evidence suggesting that LAK cells do not lyse normal haemopoietic cells, but that endothelial cells and epithelial cells from glomeruli and renal tubules are susceptible (35, 36, 42). Thus, LAK cells may damage a solid organ graft either by direct parenchymal injury or by ischaemia secondary to interruption of vascular supply.

2.4 Target cell structures recognized by NK and LAK cells

A single NK cell is capable of recognizing a heterogeneous collection of target structures (TS) expressed on tumours, non-malignant cells at early stages of differentiation, and cells infected by intracellular microbes (6). The physicochemical nature of these target structures remains elusive, and as yet, there are no reports of TS having being cloned. In the context of transplantation, the relationship between MHC antigens and TS is of interest. Although NK cell-mediated cytotoxicity is not MHC-restricted, it is a general observation that susceptibility to lysis by NK cells correlates inversely with the density of MHC Class I antigens on the target cell surface (43, 44). The best explanation for this relationship is that MHC Class I molecules or peptides bound to them associate with, and mask TS from recognition and lysis by NK cells. All Class I molecules surveyed so far, with the exception of HLA-A2, show comparable protective abilities, whereas Class II MHC antigens have no effect on TS. Recent transfection studies using sequenced MHC genes suggest that their peptide-binding groove may be the site of interaction with TS (45).

Less is known about the target structures which are recognized by LAK cells, but it has been reported that MHC antigens do not protect TS from LAK activity, possibly because of the higher cytotoxic activity of these cells (44). Although it has been reported that the TCR of some T-cell-derived LAK cells may participate in their non-specific cytotoxicity, there is no evidence available to suggest that these cells can recognize MHC antigens as TS.

Taken together, these findings are consistent with a role in immuno-surveillance for NK cells where transformed or infected cells expressing reduced amounts of MHC antigen are more susceptible to NK activity. The close association between TS and MHC antigens may have little relevance to transplantation, except for the possibility that up-regulation of MHC antigens by inflammatory mediators might protect normally susceptible targets against injury by NK cells.

2.5 The effect of transplantation on host NK cell activity

Investigation of NK cell numbers and activity after transplantation allows comparison of the effectiveness of immunosuppression regimes and also provides data pertinent to the contribution of NK cells to transplant complications. Haematological malignancies and aplastic anaemias requiring transplantation frequently have mild to moderate impairment of NK activity, but after T-cell-depleted or unmodified bone marrow grafts, peripheral blood NK activity returns to normal levels within 3 to 4 weeks (5, 46–51). At this time in the recovery of bone marrow function, a large proportion of PBL are activated $CD16^+$, $CD56^+$ NK cells or LAK cells with enhanced cytotoxic potential (41, 47, 49–51). Cytogenetic analysis of cloned NK cells from one patient after transplantation indicated a donor origin, but prompt recovery of NK cell function following grafts largely depleted of NK cells suggests a recipient contribution (5, 47). Recipient peripheral blood NK activity is unaffected or even enhanced by split dose irradiation before transplantation, but with the addition of cyclophosphamide therapy, NK cell activity remains unmeasurable until low PBL counts recover with successful engraftment (46, 49–51). The IL-2-inducible precursors of LAK cells also reappear early after transplantation and are mainly of the $CD56^+$ NK cell phenotype (49, 51). The early reappearance of functional NK and LAK cells may be advantageous to the recipient, as several months may elapse before the complete recovery of B- and T-cell function after bone marrow transplantation.

NK cell activity in recipients of solid organ grafts remains impaired for much longer periods of time due to ongoing immunosuppression. Corticosteroid therapy has only minor suppressive effects on NK cell function, but antiproliferative drugs, such as cyclophosphamide and azathioprine, deplete NK precursor cells, producing severe and prolonged reductions in spontaneous and IFN-inducible NK activity against virus-infected and malignant target cells (50, 54, 55). In renal allograft recipients, full recovery of NK cell function after cessation of azathioprine may take up to 3 months (52). Inhibition of IL-2 production by Cyclosporine A results in only mild reductions in NK cell numbers without any direct effect on the cytotoxic activity of NK cells (54, 55).

3   Bone marrow transplantation

3.1   The role of NK cells in experimental bone marrow graft rejection: characterization of hybrid and allogeneic resistance

One of the major challenges in clinical bone marrow transplantation is elucidation of the mechanisms responsible for the 2–25 per cent incidence of graft failure complicating MHC-identical sibling grafts (46, 56, 57) (see Chapter 2). Non-T-cell-mediated cytotoxic mechanisms, such as NK activity, may eliminate donor haemopoietic stem cells, as specific T-cell-mediated rejection should only be recruited by MHC disparity. It is increasingly apparent that a proportion of rejection events may still be T-cell-mediated reactions provoked by incompatibilities at tissue-specific antigen or mH antigen loci, many of which are not MHC-linked (57–59). The mH antigens so far identified are the male chromosome-associated Y antigen and five other antigens which are differentially expressed on lymphoid and haemopoietic stem cells as well as other cell types, such as keratinocytes (59). These antigens do not provoke strong antibody responses and can only be detected *in vitro* by primed T-cell responses in previously sensitized individuals (57). One identified non-cytotoxic mechanism predisposing the recipient to graft failure is the use of T-cell-depleted grafts, which diminishes GVHD but deprives donor stem cells of colony stimulating factor (CSF) support (46).

Evidence accumulated over the last 25 years from the experimental model of hybrid resistance (HR) strongly supports a role for NK cells in non-MHC-determined bone marrow graft (BMG) rejection (60, 61). Hybrid resistance was first clearly demonstrated when BMGs from selected different inbred murine strains carrying identical haplotypes at both H-2 loci were promptly rejected by their F1 hybrid offspring (Fig. 5.1A). Rejection was limited to haematopoietic stem cells and lymphoid cells, as grafts of skin and other organs were accepted without any signs of immunological recognition (60–63). In the standard experimental model of HR, recipient animals are lethally irradiated to destroy haematopoietic stem cells, and recolonization by donor stem cells is assayed by splenic colony or proliferation assays. Because F1 hybrids inherited one MHC haplotype from each parent, the traditional view was that HR was not mediated by F1 T-cells which would not recognize either parental MHC phenotype as foreign. Hybrid resistance was also not a non-specific suppressive effect of recipient NK cells, as syngeneic BMG were not suppressed *in vivo* (63, 64). Ample evidence also exists to refute the theory that HR was primarily due to ADCC-mediated by recipient NK cells directed against the BMG by anti-stem cell 'natural antibodies' of recipient origin (65).

Although T-cell reactivity against MHC antigens could not be implicated in HR, the possibility remained that F1 hybrid T-cells recognized mH antigens; especially as limited *in vitro* evidence suggested that F1 hybrid T-cells cultured with parental splenocytes gained antiparental reactivity (66). However, the effector cells responsible for HR were proven unequivocally to be NK cells by: (1) abrogation of HR *in vivo* by pretreatment with NK cell-specific antisera, oestrogen, cyclophosphamide, ^{89}Sr or split dose irradiation; (2) restoration of HR in NK cell-depleted mice by adoptive transfer of pure NK cell populations; (3) rapid boosting of HR by IFNα and IFN inducers; and (4) inability to mediate HR in mice naturally lacking NK cells due to the beige mutation (bg/bg) or immaturity (60–64, 67). Similar depletion and transfer experiments using T-cells provided strong

158   *P. G. Hogan*

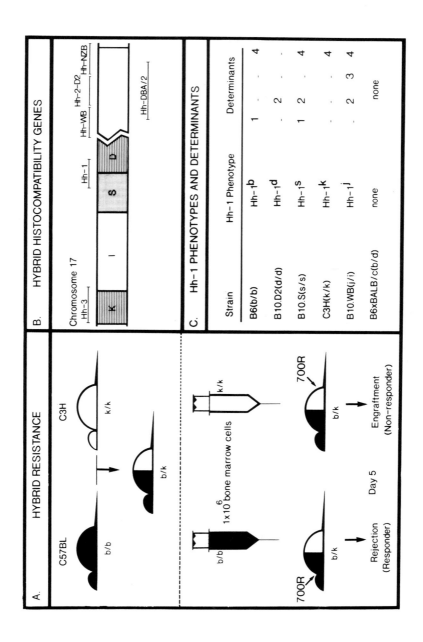

evidence against a role for specific T-cell-mediated immunity in HR. Further confirmation came from studies of NK-competent mice with severe combined immunodeficiency (scid), where HR/AR was intact or paradoxically increased (68).

The phenomenon of HR is unequivocally restricted to certain combinations of donor and recipient H-2 phenotypes. Consistent demonstrations of HR towards parental murine strains B6 ($H-2^b$), B10.D2 ($H-2^d$), B10.WB ($H-2^j$) C3H ($H-2^k$) and B10.S ($H-2^s$) have now been reported; the common theme being that only H-2-homozygous BMG cells were susceptible to HR by haploidentical F1 offspring (60–64). The limitations of the HR model include: (1) a dose-dependent effect, where increasing numbers of bone marrow cells eventually override HR; (2) the development of GVHD, which can suppress HR; (3) variable experimental requirements for different mouse strains; (4) use of the anti-asialo-GM1 antiserum which identifies activated T-cells as well as NK cells; and (5) the use of a splenic assay which measures haemopoietic engraftment in an extramedullary site (60, 69). In addition, conditioning therapy with single dose irradiation is intended to ablate recipient T-cells and spare NK cells, but a minority of radioresistant T-cells survive with intact function in some assay systems (70). Nevertheless, with adherence to uniform conditions, the experimental model of HR produces consistent results despite various limitations. Allogeneic resistance (AR) is a cellular mechanism of allogeneic BMG rejection functionally identical to HR (63, 71–74). Transplantation of certain H-2 homozygous BMGs into allogeneic irradiated recipients may, therefore, provoke antigraft reactions by both MHC antigen-specific radioresistant T-cells and recipient NK cells mediating AR. In the case of donor B6 strain mice ($H-2^b$), AR to a B6 bone marrow allograft was noted

---

Fig. 5.1 (A) Hybrid resistance of the C57BL × C3H F1 hybrid ($H-2^b/H-2^k$) against bone marrow transplants from parental strains. After pretransplant conditioning with radiotherapy to ablate host haematopoietic stem cells, 1×10^6 bone marrow cells from either parent were infused. Grafts from C57BL parents were rejected on day 5 as assayed by a spleen colony assay. Grafts from C3H parents survived because the C57BL × C3H hybrid is a nonresponder to $H-2^k$ grafts. (Adapted from (61).) (B) The location of hybrid histocompatibility (Hh) genes on the mouse chromosome 17 in relation to the H-2 gene complex. The Hh-1 gene is responsible for both allogeneic and hybrid resistance and is located between H-2S (Class III MHC genes) and H-2D (Class I MHC genes). Other Hh gene loci designated according to the strain in which HR is mediated map away from H-2 in the relative positions shown, except for Hh-DBA/2 which is located on another chromosome. Hh-3 is located at H-2K (Class I) and mediates AR only. H-2I represents Class II MHC genes. (C) The expression of Hh-1 determinants expressed by selected mouse strains of different Hh-1 phenotypes. The Hh-1 determinants were numbered arbitrarily according to their order of detection. (Adapted from (77, 78).)

in CB.17 (H-2^d), DBA/2 (H-2^d) and C3D2F1 (H-2^k)/H-2^d) recipients, although in the case of CB.17 the resistance was relatively weak. After allogeneic transplantation, therefore, BMG failure may occur by dual mechanisms; AR mediated by NK cells commences by 12 h and persists for up to 4 days, whereas T-cell reactivity surviving irradiation commences at day 7 (74). In one model of AR, a single lineage of cells expressing both NK markers and T-cell receptors has been proposed as responsible for AR, but the data is also consistent with separate effector cell populations (75). It is probable that the mechanism of AR accounts for the independently described phenomenon of natural cytotoxicity or allogeneic lymphocyte cytotoxicity where infused lymphocytes are rapidly cleared from the circulation by the lymphoid organs of allogeneic animals (8). If preformed donor-specific antibodies are present, ADCC mediated by NK cells may also contribute to AR.

Within a particular H-2 phenotype, the ability to mediate HR/AR is variable and appears to be a dominantly inherited, non-H-2-linked trait. In the case of 'poor responder' strains, such as CB.20 (H-2^d), the defect can be corrected by therapy with IFNα, which stimulates mature NK cells and recruits their precursors (60, 63, 67). Thus, the outcome of a H-2 homozygous BMG is determined not only by the H-2 phenotype of the recipient but also by an independently inherited responder status. Many facets of the NK cell-mediated mechanism underlying HR and AR remain to be elucidated. In particular, the need for a prior induction phase and the nature and control of the target structure(s) on haemopoietic stem cells are two features which appear novel to the NK cell system mediating HR/AR. Data from kinetic and IFN administration studies suggest that an induction phase involving secretion of IFN is necessary to support the development of HR/AR (67). Silica-sensitive cells, such as recipient macrophages, are the probable producer cells of IFN; T-cells from either donor or recipient are less likely candidates because of the facility with which scid mice participate in HR/AR (67, 68). The nature and source of the stimulus for IFN secretion by macrophages is speculative, but it is possible that NK cells may themselves provide the stimulus after interaction with target structures on stem cells. A similar need for IFN secretion by accessory cells has been demonstrated for the lysis of herpes simplex virus-infected tumour cells by human peripheral blood NK cells (76).

Present concepts of the target recognition system enabling NK cells to recognize and respond only to H-2 homozygous BMG stem cells rely largely on experimental studies which infer the existence of a family of murine hybrid histocompatibility (Hh) genes located with one known exception, on chromosome 17 in proximity to the H-2 genes (60, 63). The Hh antigen phenotype of a H-2 homozygous strain is determined by the following procedures: (1) BMG from strains with similar Hh antigens will show the same profile of rejection or acceptance when transferred into a panel of hosts; (2) F1 hybrids of Hh-identical strains will not reject a BMG from either parent, whereas quantitatively different degrees of engraftment indicate differing parental Hh profiles (60, 77, 78). The first Hh antigen described, designated arbitarily as Hh-1, determines patterns of HR and AR for the bulk of murine strains (Fig. 5.1B). Studies of intra-H-2 recombinant inbred strains show that the Hh-1 gene is not a Class I MHC gene and is situated between the H-2S and H-2D regions (77, 78). The remaining four Hh genes have weaker effects on HR/AR and are limited in their effect to only a few or single donor strains (60, 77, 78). Cross-breeding and transplantation studies show considerable polymorphism among Hh-1$^+$ strains, with the expression of up to four determinants controlled by this locus. The resulting phenotype of a particular strain can generally be predicted according to H-2D type, although in some strains recombination events adjacent to or within Hh-1 lead to unexpected Hh-1 antigen phenotypes (Fig. 5.1C).

Based on this data, the model of HR/AR implies that recipient NK cells are capable of rejecting normal H-2 homozygous donor cells expressing non-recipient Hh determinants. However, in the context of the Hh system, there is no plausible molecular model explaining how non-recipient combinations of H-1 determinants are perceived as foreign by recipient NK cells. If Hh genes coded directly for a membrane NK TS, a recessive form of inheritance would be consistent with the need for HR and AR-susceptible strains to be H-2 homozygous. Alternatively, Hh genes may be regulatory, exerting codominant *trans*-acting suppressive effects on NK-susceptible membrane TS which would otherwise be fully expressed constitutively. Thus, Hh-heterozygous animals with a complete or near-complete complement of suppressive genes would lack NK cell-sensitive TS, whereas homozygous animals missing some genes would retain TS and sensitivity to HR/AR. (Fig. 5.1C). A recent molecular approach towards the testing of Hh-1 gene function used B6 (H-2^b) mice transgenic for a H-2D^d gene sequence(79). Although Hh genes are not located in the H-2D area, normally susceptible B6 donor BMGs were protected against HR mediated by H-2^b/H-2^d F1 hybrids by the presence of the transgene. This unconfirmed finding is inconsistent with the proposed role of the Hh-1 gene system in HR, but it is possible that insertion of the transgene interfered non-specifically with the structural or regulatory function of nearby Hh-1 genes.

In addition to their participation in HR/AR, NK cells may suppress haemopoietic colony formation, not only by direct cytotoxicity but also by the secretion of colony-inhibiting factors. Conflicting results have arisen from the use of different bone marrow cell preparations in cocultures, with some investigators reporting that NK cells have no detectable effect on purified stem cells (80). However, the majority of studies on whole bone marrow populations report that NK cells can exert a suppressive effect.

Following brief contact with autologous or allogeneic bone marrow, NK cells produce soluble cytostatic factors which retard the development of most haemopoietic lineages (81). The exact identity of the colony inhibitory activity is undecided but TNFα and IFNγ may be involved. There are a few reports that NK cells elaborate CSFs (82), but clinical descriptions of large granular lymphocytic leukaemias associated with neutropaenia suggest that *in vivo*, NK cells are likely to suppress granulopoiesis (83).

Although the phenomena of HR and AR provide compelling evidence for NK cell-mediated mechanisms of experimental BMG rejection, there is a distinct lack of data supporting a similar role for NK cells in human bone marrow transplantation. The entire HR/AR model may be a clinically irrelevant consequence of the genetic homogeneity of inbred murine strains, although it is possible to reproduce the effect in outbred animals, such as dogs (60). The only relevant data providing limited support for the occurrence of HR/AR in clinical transplantation derive from studies of congenital immunodeficiency syndromes and NK activity after transplantation. Transplantation of scid patients with genotypically identical marrow is usually successful, but mismatched cases often fail to engraft, particularly when the recipient has high levels of NK activity prior to transplantation (84). In the case of leucocyte adhesion deficiency, the consistent success of MHC-non-identical grafts may be attributable in part to defective NK activity, although such patients also have depressed T-cell function (85). Serial assays of peripheral blood NK activity and phenotype after MHC-identical bone marrow transplants argue against a role for NK cells in rejection. After successful engraftment, NK activity is present at normal levels, but during graft rejection, NK activity is often low or undetectable (5). Although this data is used as evidence against the participation of NK cells, another interpretation may be that circulating NK cells have migrated to the bone marrow to mediate rejection.

3.2   Graft-versus-host-disease

Acute and chronic GVHD are major complications of BMT which result in severe cutaneous, gastrointestinal and pulmonary morbidity and mortality in a large proportion of recipients (46). It is difficult to apply experimental and clinical NK cell data to the pathogenesis of acute and chronic GVHD because they are separate but overlapping clinical syndromes which may be caused by different immunological mechanisms. Acute GVHD, developing within 100 days after transplantation, is characterized by the presence of host-reactive donor cytotoxic T-cells and a high incidence of opportunistic infection, whereas hallmark features of the chronic form are autoimmunity and high levels of non-specific suppressor cell activity (46).

Although donor cytotoxic T-cells reactive against recipient MHC or mH antigens are established as a major cause of tissue damage in GVHD, many of the clinical and pathological features cannot be explained solely on this basis. Host-reactive T-cells of donor origin can be detected in recipients irrespective of the presence of GVHD; raising the possibility of tolerance mediated by suppressor cells or the participation of other cytotoxic cells, such as NK cells *in vivo* (86).

Clinical observations relating elevated pre-or post-transplant peripheral blood NK activity to the occurrence and severity of GVHD provide indirect evidence of dubious significance in support of an NK cell-mediated component of GVHD (87, 88). These studies have been contested by other authors using the same targets, and in any case, the data may be explained by a non-specific elevation of host NK activity as a result of GVHD-related immune stimulation (48, 50, 89). In common with many other studies of NK cells, these results are dependent on the target cell type, as NK activity against CMV-infected fibroblasts and HR against bone marrow cells are suppressed during GVHD (69, 90). Finally, ultrastructural studies demonstrated lymphocyte-associated epithelial damage in patients with GVHD, but there is very little immunohistological indication that the effectors are NK cells (91, 92).

A large number of experimental studies have used NK cell depletion techniques, immunohistology, and NK-deficient beige mice to provide indisputable evidence that donor NK cells have an active cytotoxic or accessory role in experimental GVHD.

In immunocompetent animals, abnormally high host NK activity in peripheral blood or lung interstitium after transplantation bears no relation to the presence or degree of organ damage in GVHD, and as such, is probably an epiphenomenon (93). On the other hand, donor NK cells appear to contribute directly to the pathological lesions characteristic of GVHD. Large granular lymphocytes (LGLs) of donor origin with NK-specific surface antigens congregate adjacent to degenerating epithelial cells in the liver, skin, and intestines of mice undergoing acute GVHD (94). Studies where either donor or host NK cells are depleted pretransplantation provide confirmation that NK cells of donor origin are the important effectors of GVHD injury. In models of GVHD provoked by mH antigen disparity; beige (bg/bg) donor marrow transfer was associated with splenomegaly, immunosuppression of B-and T-cell functions and a lack of organ pathology, whereas full-blown GVHD resulted from NK-competent (bg/+) donors (95). Treatment of prospective donors with NK-specific antiserum achieves the same effect, provided that NK precursor cells are also made susceptible to depletion by allogeneic induction prior to transplantation (96). If mature donor NK cells alone are depleted, NK precursor cells can proliferate in the recipient, resulting in GVHD (96). This fact explains why some authors have reported that donor NK cell depletion does not influence the subsequent course of GVHD (97). Similarly, reports that recipient NK cells play a stimulatory role are based on depletion studies which would also neutralize transplanted donor NK cells and their precursors (98). As yet, no study has specifically addressed this issue by using recipients with isolated NK cell defects. However, it was noted that there was no difference in the severity of GVHD in bg/bg or bg/ + recipients of bg/+ bone marrow (95). Thus, the experimental evidence favours the hypothesis that GVHD is initiated by activation of donor alloreactive T-cells which, in turn, activate donor NK cells and their precursors by the secretion of IL-2 and IFNγ. The observation that only donor NK cells may be injurious to host organs raises the possibility that Hh antigens may be the TS.

3.3　The protective role of NK cells after bone marrow transplantation

After successful engraftment, recurrence of leukaemia or opportunistic infections are the major causes of mortality in bone marrow recipients (46). The rapid recovery of NK cells after transplantation suggests that they should be an important component of host resistance to these complications, particularly when specific immune defences are still recovering from conditioning with radiotherapy and chemotherapy. A lack of association between recipient NK activity assayed against K562 targets and the incidence of leukaemia after transplantation has been reported, but data derived from irrelevant targets may be misleading (48). A comprehensive post transplantation study using CMV-infected, allogeneic fibroblasts as NK targets detected an inverse correlation between serial NK activity and the incidence and severity of CMV infection (90). Clinical observations that patients with GVHD had lower leukaemia relapse rates lead to the concept of a graft-versus-leukaemia effect where activated cytotoxic cells or cytokines from the graft eliminated leukaemic cells which survived transplantation. There is currently doubt as to whether GVHD is required for induction of the mechanism and which cells are the major effectors (46). In clinical studies of NK cells after transplantation, there are no prospective data relating NK activity to leukaemia relapse, but the sensitivity of leukaemic blasts to LAK activity *in vitro* suggests that LAK activity after transplantation may retard the emergence of leukaemic clones (47–49). Stimulating bone marrow with IL-2 prior to experimental transplantation produces a marked graft-versus-tumour effect *in vivo* without a deleterious effect on engraftment (99). The antileukaemic activity of LAK cells has also been utilized to purge minor leukaemic populations from the bone marrow *in vitro*, without any apparent stem cell damage (42). In the future, the incidence of malignancy relapse may be reduced further by treatment of the graft cells prior to infusion by LAK cell purging or stimulation of graft LAK cells with IL-2.

4   Solid organ transplantation

4.1   The contribution of NK and LAK cells to solid organ graft rejection

The evidence implicating NK or LAK cells in the rejection of solid organ allografts is largely indirect and less convincing than the data for bone marrow transplantation. Graft-infiltrating lymphocytes displaying NK cell surface antigens are certainly detected in increased numbers together with T-cells, macrophages, and granulocytes during rejection, but it remains questionable whether their cytotoxic, accessory or pro-inflammatory functions have any bearing on graft survival (100–103). Clinical and experimental studies reporting no association between rejection and NK cell numbers or activity in either peripheral blood or graft are not unexpected as NK cells are incapable of lysing graft tissue (104, 105). However, the role of LAK cells warrants investigation in rejection because conditions supporting the generation of LAK cells exist within grafts. During rejection, the presence of IL-2 transcripts and LAK precursur cells in graft tissue indicates the potential for generation of high level LAK activity towards endothelial and renal parenchymal cells (35, 39, 106–108). In an experimental model of canine lung transplant rejection, reduced pulmonary perfusion was associated with bronchopulmonary lavage LAK activity which preceded the appearance of allospecific T-cells (106). As yet, there are no similar data available from clinical studies of allograft rejection.

5   Summary

NK and LAK cells may be the first effector cells to encounter foreign graft tissue and, as such, deserve further investigation aimed specifically at their function early after transplantation. The proposition that NK and LAK cells play a significant role in the many facets of transplantation is based substantially on animal experimentation. Although much of this data has no direct relevance to clinical practice, it indicates potentially fruitful directions for future research. In bone marrow transplantation, the HR/AR models point towards a role for NK cells in the early phases of rejection, particularly when grafts are genotypically mismatched. Identification and cloning of the stem cell TS which triggers NK cells during HR could lead to the discovery of similar structures on human bone marrow cells; thereby providing a new approach for defining the clinically significant effects of NK cells on haemopoiesis. The knowledge that NK cells respond to specific targets in the models of HR and AR suggests that NK or LAK cells may play an active role in responding to transplated tissue rather than being passively attracted to the graft by cytokines secreted by specific T-cells.

The strong possibility that NK and LAK cells are important protective barriers against infection and malignancy early after transplantation suggests that one aim of bone marrow graft modification techniques and immunosuppressive therapy should be to optimize levels of non-specific cytotoxicity in the recipient after grafting. Although animal models indicate that this approach may exacerbate GVHD, there is no evidence that GVHD and NK activity are strongly linked in humans.

The overall contribution of LAK cells to the rejection of solid organ allografts is probably minimal in comparison to the effect of specific T-cells. Nevertheless, the ability of cytotoxic LAK cells to infiltrate grafts before the appearance of the alloantigen-specific mechanisms suggests that their function in this setting may be more than merely that of a bystander.

Acknowledgements

This review was supported in part by the Australian National Council On Aids (Department of Community Services and Health), and the Lions Kidney and Medical Research Foundation. Critical review of this manuscript by Associate Professor Ian Frazer and preparation of the figure by Mrs S. Healey are gratefully acknowledged.

References

1 Halloran, P. F., Cockfield, S. M., and Madrenas, J. (1989). The molecular immunology of transplantation and graft rejection. *Immunol. All. Clin. North Am.*, **9**, 1–19.

2 Loveland, B. and Simpson, E. (1986). The non-MHC transplantation antigens: neither weak or minor. *Immunol. Today.*, **7**, 223–9.

3 Eckels, D. (1990). Alloreactivity: Allogeneic presentation of endogenous peptide or direct recognition of MHC polymorphism? A review. *Tiss. Antigens,* **35**, 49–55.

4 Colvin, R. B. (1990). Cellular and molecular mechanisms of allograft rejection. *Ann. Rev. Med.*, **41**, 361–75.

5 Bordignon, C., *et al.* (1989). Graft failure after T-cell-depleted human leukocyte antigen identical marrow transplants for leukaemia: II. *In vitro* analyses of host effector mechanisms. *Blood,* **74**, 2237–43.

6 Trinchieri, G. (1989). Biology of natural killer cells. *Adv. Immunol.*, **47**, 187–376.

7 Sissons, J. G. P. (1986). The immunology of cytomegalovirus infection. *J. Roy. Coll. Phys.*, **20**, 40–44.

8 Rolstad, B. and Ford, W. L. (1983). The rapid elimination of allogeneic lymphocytes: relationship to established mechanisms of immunity and to lymphocyte traffic. *Immunol. Rev.*, **73**, 87–113.

9 Timonen, T., Ortaldo, J. R., and Herberman, R. B. (1981). Characteristics of human large granular lymphocytes and relationship to natural killer and K cells. *J. Exp. Med.*, **153**, 569–82.

10 Lotzova, E. and Savary, C. A. (1987). Generation of NK cell activity from human bone marrow. *J. Immunol.*, **139**, 279–84.

11 Abo, T., Miller, C. A., and Balch, C. M. (1984). Characterization of granular lymphocyte subpopulations expressing HNK-1 (leu 7) and leu 11 antigens in the blood and lymphoid tissues from fetuses, neonates and adults. *Eur. J. Immunol.*, **14**, 616–23.

12 Weissler, J. C., Nicod, L. P., Lipscomb, M. B., and Toews, G. B. (1987). Natural killer cell function in lungs is compartmentalized. *Am. Rev. Resp. Dis.*, **135**, 941–9.

13 Hogan, P. G., Hapel, A. J., and Doe, W. F. (1986). Lymphokine-activated and natural killer cell activity in human intestinal mucosa. *J. Immunol.*, **135**, 941–9.

14 Botazza, B., Introna, M., Allavena, P., and Mantovani, A. (1985). *In vitro* migration of human large granular lymphocytes. *J. Immunol.*, **134**, 2316–21.

15 Aronson, F. R., Libby, P., Brandon, E. P., Janicka, M. W., and Mier, J. W. (1988). IL-2 rapidly induces natural killer cell adhesion to human endothelial cells. A potential mechanism for endothelial injury. *J. Immunol.*, **141**, 158–63.

16 Bender, J. R., Pardi, R., Karasek, M. A., and Engleman, E. G. (1987). Phenotypic and functional characterization of lymphocytes that bind human microvascular endothelial cells *in vitro*. *J. Clin. Invest.*, **79**, 1679–88.

17 Kohl, S., Springer, T. A., Schmalsteig, F. C., Loo, L. S., and Anderson, D. C. (1984). Defective natural killer cytotoxicity and polymorphonuclear leucocyte antibody-dependent cellular cytotoxicity in patients with LFA-1/OKM-1 deficiency. *J. Immunol.*, **133**, 2972–8.

18 Patarroyo, M., *et al.* (1990). Leukocyte-cell adhesion: a molecular process fundamental in leukocyte physiology. *Immunol. Rev.*, **114**, 67–108.

19 Holmberg, L. A., Miller, B. A., and Ault, K. A. (1984). The effect of natural killer cells on the development of syngeneic hematopoietic progenitors. *J. Immunol.*, **133**, 2933–9.

20 Miller, S. C., Galley, D., and Nguyen, D. M. (1988). Inhibition of natural killer cell-mediated lysis of tumor cells by normal and regenerating bone marrow. *Immunobiol.*, **177**, 82–90.

21 Hogan, P. G. and Basten, A. (1988). What are killer cells and what do they do? *Blood Rev.*, **2**, 50–8.

22 Scala, G. S., *et al.* (1986). Cytokine secretion and noncytotoxic functions of human large granular lymphocytes. In *Immunobiology of natural killer cells* (ed. E. Lotzova, and R. B. Herberman), CRC Press, Boca Raton, Florida.

23 Degliantoni, G., Murphy, M., Kobayashi, M., Francis, M., Perussia, B., and Trinchieri, G. (1985). Natural killer (NK) cell-derived hematopoietic colony-inhibiting activity and NK cytotoxic factor. Relationship with tumor necrosis factor and synergism with immune interferon. *J. Exp. Med.*, **162**, 1512–30.

24 Suzuki, R., Suzuki, S., Ebina, N., and Kumagai, K. (1985). Suppression of alloimmune cytotoxic T lymphocyte (CTL) generation by depletion of NK cells and restoration by interferon and/or interleukin 2. *J. Immunol.*, **134**, 2139–4823.

25 Pope, R. M., McChesney, L., Stebbing, N., Goldstein, L., and Talal, N. (1985). Regulation of T cell proliferation by cloned interferon-α mediated by Leu-11b positive cells. *J. Immunol.*, **135**, 4048–53.

26 Trinchieri, G. and Perussia, B. (1985). Immune interferon: a pleiotropic lymphokine with multiple effects. *Immunol. Today*, **6**, 131–6.

27 Bottazzo, G. F., Todd, I., Mirakian, R., Belfiore, A., and Pujol-Borrell, R. (1986). Organ-specific autoimmunity: a 1986 overview. *Immunol. Rev.*, **94**, 137–69.

28 Abruzzo, L. V. and Rowley, D. A. (1983). Homeostasis of the antibody response: immunoregulation by NK cells. *Science*, **222**, 581–5.

29 Pistoia, V., Cozzolino, F., Torcia, M., Castigli, E., and Ferrarini, M. (1985). Production of B cell growth factor by a Leu-7$^+$, OKM1$^+$ non-T cell with the features of large granular lymphocytes. *J. Immunol.*, **143**, 3179–84.

30 Scala, G., *et al.* (1984). Human large granular lymphocytes (LGL) are potent producers of interleukin 1. *Nature*, **309**, 56–9.

31 Goldfarb, R. H., Timonen, T., and Herberman, R. B. (1984). Production of plasminogen activator by natural killer cells. *J. Exp. Med.*, **159**, 935–51.

32 Scala, G., Allavena, P., Ortaldo, J. R., Herberman, R. B., and Oppenheim, J. J. (1985). Subsets of human large granular lymphocytes (LGL) exhibit accessory cell functions. *J. Immunol.*, **134**, 3049–55.

33 Herberman, R. B., Nunn, M. E., Lavrin, D. H., and Asofsky, R. (1973). Effect of antibody to θ antigen on cell-mediated immunity induced in syngeneic mice by murine sarcoma virus. *J. Nat. Cancer Inst.*, **51**, 1509–12.

34 Grimm, E. A., Mazumder, A., Zhang, H. Z., and Rosenberg, S. A. (1982). Lymphokine-activated killer cell phenomenon. Lysis of natural killer resistant fresh solid tumor cells by interleukin 2-activated autologous human peripheral blood lymphocytes. *J. Exp. Med.*, **155**, 1823–41.

35 Damle, N. K., Doyle, L. V., Bender, J. R., and Bradley, E. C. (1987). Interleukin 2-activated human lymphocytes exhibit enhanced adhesion to normal vascular endothelial cells and cause their lysis. *J. Immunol.*, **38**, 1779–85.

36 Nietosvaara, Y., Renkonen, R., Mattila, P., and Hayry, P. (1990). Cytotoxic lymphocytes and lymphokine-activated killer cell-mediated lysis against rat kidney components. *Transplant. Proc.*, **22**, 127.

37 Phillips, J. H., Le, A. M., and Lanier, L. L. (1984). Natural killer cells activated in a human mixed lymphocyte response culture identified by expression of Leu-11 and class II histocompatibility antigens. *J. Exp. Med.*, **159**, 993–1008.

38 Moingeon, P., *et al.* (1985). A target structure for a series of human cloned natural killer cell lines is recognized by both anti TNKtar and 4F2 monoclonal antibodies. *J. Immunol.*, **134**, 2930–4.

39 Lowry, R. P., Wang, K., Vernooij, B., and Harcus, D. (1989). Lymphokine transcription in vascularized mouse heart grafts: effects of 'tolerance' induction. *Transplant. proc.*, **21**, 72–3.

40 Chatenoud, L., *et al.* (1990). *In vivo* cell activation following OKT3 administration. Systemic cytokine release and modulation by corticosteroids. *Transplantation*, **49**, 697–702.

41 Rooney, C. M., Wimperis, J. Z., Brenner, M. K., Patterson, J., Hoffbrand, A. V., and Prentice, H. G. (1986). Natural killer cell activity following T cell-depleted allogeneic bone marrow transplantation. *Brit. J. Haematol.*, **62**, 413–20.

42 Van Den Brink, M. R. M., Voogt, P. J., Marijt, W. A. F., Van Luxemburg-Heijs, S. A. P., Van Rood, J. J., and Brand, A. (1989). Lymphokine-activated killer cells selectively kill tumour cells in bone marrow without compromising bone marrow stem cell funtion *in vitro*. *Bone Marrow Transplant.*, **4** (Suppl. 2), 72A.

43 Storkus, W. J., Howell, D. N., Salter, R. D., Dawson, J. R., and Cresswell, P. (1987). NK susceptibility varies inversely with target cell class I HLA antigen expression. *J. Immunol.*, **138**, 1657–9.

44 Storkus, W. J., Alexander, J., Payne, A., Dawson, J. R., and Cresswell, P. (1989). Reversal of natural killing susceptibility in target cells expressing transfected class I HLA genes. *Proc. Natl. Acad. U.S.A.*, **86**, 2361–4.

45 Storkus, W. J., Alexander, J., Payne, J. A., Creswell, P., and Dawson, J. R. (1989). The α1/α2 domains of class I HLA molecules confer resistance to natural killing. *J. Immunol.*, **143**, 3853–7.

46 Martin, P. J., Hansen, J. A., Storb, R., and Donnall Thomas, E. (1987). Human marrow transplantation: an immunological perspective. *Adv. Immunol.*, **40**, 379–438.

47 Hercend, T., *et al.* (1986). Characterization of natural killer cells with antileukaemia activity following allogeneic bone marrow transplantation. *Blood*, **67**, 722–8.

48 Livnat, S., Seigneuret, M., Storb, R., and Prentice, R. (1980). Analysis of cytotoxic effector cell function in patients with leukaemia or aplastic anaemia before and after bone marrow transplantation. *J. Immunol.*, **124**, 481–90.

49 Higuchi, C. M., Thompson, J. A., Cox, T., Lindgren, C. G., Buckner, C. D., and Fefer, A. (1989). Lymphokine-activated killer function following autologous bone marrow transplantation for refractory hematological malignancies. *Cancer Res.*, **49**, 5509–13.

50 Hokland, M., Jacobsen, N., Ellegaard, J., and Hokland, P. (1988). Natural killer function following allogeneic bone marrow transplantation. Very early reemergence but strong dependence of cytomegalovirus infection. *Transplantation*, **45**, 1080–4.

51 Keever, C. A., et al. (1987). Interleukin 2-activated killer cells in patients following transplants of soybean lectin-separated and E rosette-depleted bone marrow. *Blood*, **70**, 1893–1903.

52 Katz, P., Zaytoun, A. M., and Lee, J. H. (1984). The effects of *in vivo* hydrocortisone on lymphocyte-mediated cytotoxicity. *Arth. Rheum.*, **27**, 72–8.

53 Spina, C. (1984). Azathioprene as an immune modulating drug: clinical implications. *Clinics Immunol. Allergy*, **41**, 415–46.

54 Versluis, D. J., Metselaar, H. J., Bijma, A. M., Vaessen, L. M. B., Wenting, G. J., and Weimar, W. (1988). The effect of long term cyclosporine therapy on natural killer cell activity. *Transplant. Proc.*, **20**(Suppl. 2), 179–85.

55 Cauda, R., Citterio, F., Tamburrini, E., Pozzetto, U., Tumbarello, M., Castagneto, M., and Ortona, L. (1988). Natural killer activity against viral targets in cyclosporine-treated renal allograft recipients. *Transplant. Proc.*, **20**(Suppl. 2), 186–9.

56 Deeg, H. J., et al. (1986). Decreased incidence of marrow graft rejection in patients with severe aplastic anaemia: changing impact of risk factors. *Blood*, **68**, 1363–8.

57 Voogt, P. J., et al. (1988). Cellularly defined minor histocompatibility antigens are differentially expressed on human hematopoietic progenitor cells. *J. Exp. Med.*, **168**, 2337–47.

58 Morris, P. J., Fuggle, S. V., Ting, A., and Wood, K. J. (1987). HLA and organ transplantation. *Br. Med. Bull.*, **43**, 184–202.

59 Goulmy, E., Van Els, C., de Bueger, M., Kempenaar, J., Ponec, M., van Rood, J. J. (1989). Evidence for minor histocompatibility antigen expression in human skin. *Bone Marrow Transplant.*, **4**(Suppl. 4), 117A.

60 Bennett, M. (1987). Biology and genetics of hybrid resistance. *Adv. Immunol.*, **41**, 333–445.

61 Cudkowicz, G., and Stimpfling, J. H. (1964). Deficient growth of C57BL marrow cells transplanted in F_1 hybrid mice. *Immunology*, **7**, 291–306.

62 Cudkowicz, G. and Bennett, M. (1971). Peculiar immunobiology of bone marrow allografts. II. Rejection of parental grafts by resistant F1 hybrid mice. *J. Exp. Med.*, **134**, 1513–28.

63 Cudkowicz, G. and Bennett, M. (1971). Peculiar immunobiology of bone marrow allografts. I. Graft rejection by irradiated responder mice. *J. Exp. Med.*, **134**, 83–102.

64 Bordignon, C., Daley, J. P., and Nakamura, I. (1985). Hematopoietic histoincompatibility reactions by NK cells *in vitro*: model for genetic resistance to marrow cells. *Science*, **230**, 1398–1401.

65 Warner, J. F. and Dennert, G. (1985). Bone marrow graft rejection as a function of antibody-directed natural killer cells. *J. Exp. Med.*, **161**, 563–76.

66 Nakano, K., Nakamura, I., and Cudkowicz, G. (1981). Generation of F1 hybrid cytotoxic T lymphocytes specific for self H-2. *Science*, **289**, 559–63.

67 Afifi, M. S., Kumar, V., and Bennett, M. (1985). Stimulation of genetic resistance to marrow grafts in mice by interferon-α/β. *J. Immunol.*, **134**, 3739–45.

68 Murphy, W. J., Kumar, V., and Bennett, M. (1987). Rejection of bone marrow allografts by mice with severe combined immune deficiency (SCID). Evidence that natural killer cells can mediate the specificity of marrow graft rejection. *J. Exp. Med.*, **165**, 1212–7.

69 Hakim, F. T., and Shearer, G. M. (1986). Abrogation of hybrid resistance to bone marrow engraftment by graft-vs-host-induced immune deficiency. *J. Immunol.*, **137**, 3109–16.

70 Aizawa, S., Toshihiko, S., Kamisaku, H., and Kubo, E. (1980). Cellular basis of the immunohematologic defects observed in short-term semi-allogeneic B6C3F1—C3H chimeras: evidence for host-versus-graft reaction initiated by radioresistant T cells. *Cell. Immunol.*, **56**, 47–57.

71 Lotzova, E., Savary, C. A., and Pollack, S. B. (1983). Prevention of rejection of allogeneic bone marrow transplants by NK 1.1 antiserum. *Transplantation*, **35**, 490–4.

72 Warner, J. F. and Dennert, G. (1982). Effects of a cloned cell line with NK activity on bone marrow transplants, tumour development and metastasis *in vivo*. *Nature*, **300**, 31–4.

73 Dennert, G., Anderson, C. G., and Warner, J. (1985). T killer cells play a role in allogeneic bone marrow graft rejection but not in hybrid resistance. *J. Immunol.*, **135**, 3729–34.

74 Murphy, W. J., Kumar, V., and Bennett, M. (1987). Acute rejection of murine bone marrow allografts by natural killer cells and T cells. Differences in kinetics and target antigens recognized. *J. Exp. Med.*, **166**, 1499–1509.

75 Yankelevich, B., Knobloch, C., Nowicki, M., and Dennert, G. 1989). A novel cell type responsible for marrow graft rejection in mice. T cells

with NK phenotype cause acute rejection of marrow graft. *J. Immunol.*, **142**, 3423–30.

76 Paya, C. V., Schoon, R. A., and Leibson, P. J. (1990). Alternative mechanisms of natural killer cell activation during Herpes Simplex virus infection. *J. Immunol.*, **144**, 4370–5.

77 Cudkowicz, G. and Nakamura, I. (1983). Genetics of the murine Hemopoietic-histocompatibility system: an overview. *Transplant. Proc.*, **15**, 2058–63.

78 Bennett, M., *et al.* (1987). Rejection of bone marrow cells by irradiated mice: NK and T cells recognize different antigens. *Transplant. Proc.*, **19**(Suppl. 7), 5–11.79.

79 Ohlen, C., *et al.* (1989). Prevention of allogeneic bone marrow graft rejection by H-2 transgene in donor mice. *Science*, **246**, 666–8.

80 Niemeyer, C. M., Sieff, C. A., Smith, B. R., Ault, K. A., and Nathan, D. G. (1989). Hematopoiesis in vitro coexists with natural killer lymphocytes. *Blood*, **74**, 2376–82.

81 Degliantoni, G., Perussia, B., Mangoni, L., and Trinchieri, G. (1985). Inhibition of bone marrow colony formation by human natural killer cells and by natural killer cell-derived colony-inhibiting activity. *J. Exp. Med.*, **161**, 1152–68.

82 Kasahara, T., Djeu, J. Y., Dougherty, S. F., and Oppenheim, J. J. (1983). Capacity of human large granular lymphocytes (LGL) to produce multiple lymphokines: interleukin 2, interferon and colony stimulating factor. *J. Immunol.*, **131**, 2379–85.

83 Grillot-courvalin, C., Vinci, G., Tsapsis, A., Dokhelar, M. C., Vainchenker, W., and Brouet, J. C. (1986). The syndrome of T8 hyperlymphocytosis: variation in phenotype and cytotoxic activities of granular cells and evaluation of their role in associated neutropenia. *Blood*, **69**, 1204–10.

84 Neudorf, S. M. L., Yanik, G. A., and Pietryza, D. W. (1990). Bone marrow transplantation for correction of primary immune deficiencies. In *Bone marrow transplantation in children* (ed. F. L. Johnson and C. Pochedly). Raven Press, New York.

85 Le Deist, F., *et al.* (1989). Successful HLA nonidentical bone marrow transplantation in three patients with the leukocyte adhesion deficiency. *Blood*, **74**, 512–6.

86 Van Els, C. A., Bakker, A., Zwinderman, A. H. Zwaan, F. E., Van Rood, J. J., and Goulmy, E. (1990). Effector mechanisms in graft-versus-host disease in response to minor histcompatibility antigens. I. Absence of correlation with cytotoxic effector cells. *Transplantation*, **50**, 62–6.

87 Lopez, C., Kirkpatrick, D., Sorell, M., O'Reilly, R. J., and Ching, C. (1979). Association between pre-transplant natural kill and graft-versus-host disease after stem cell transplantation. *Lancet*, **24**, 1103–6.

88 Dokhelar, M. C., *et al.* (1981). Natural killer cell activity in human bone marrow recipients. Early reappearance of peripheral natural killer cell activity in graft-versus-host disease. *Transplantation*, **31**, 61–5.

89 Gratama, J. W., *et al.* (1985). Natural immunity and graft-vs-host disease. *Transplantation*, **40**, 256–60.

90 Bowden, R. A., Day, L. M., Amos, D. E., and Meyers, J. D. (1986). Natural cytotoxic activity against cytomegalovirus-infected target cells following marrow transplantation. *Transplantation*, **44**, 504–8.

91 Gallucci, B. B. Sale, G. E., McDonald, G. B., Epstein, R., Shulman, H. M., and Thomas, E. D. (1982). The fine structure of human rectal epithelium in acute graft-versus-host disease. *Am. J. Surg. Pathol.*, **6**, 293–305.

92 Murphy, G. F., Merot, Y., Tong, A. K., and Smith, B. (1985). Identification of distinctive lymphocyte subpopulations in cutaneous graft-versus-host disease. *Lab. Invest.*, **52**, 46A.

93 Gartner, J. G., Merry, A. C., and Smith, C. I (1988). An analysis of pulmonary natural killer cell activity in F1-hybrid mice with acute graft-versus-host reactions. *Transplantation*, **46**, 879–86.

94 Ferrara, J. L., Guillen, F. J., Van Dijken, P. J., Marion, A., Murphy, G. F., and Burakoff, S. J. (1989). Evidence that large granular lymphocytes of donor origin mediate acute graft-versus-host disease. *Transplantation*, **47**, 50–4.

95 Ghayur, T., Seemayer, T. A., Kongshavn, P. A., Gartner, J. G., and Lapp, W. S. (1987). Graft-versus-host reactions in the beige mouse. An investigation of the role of host and donor natural killer cells in the pathogenesis of graft-versus-host disease. *Transplantation*, **44**, 261–7.

96 Ghayur, T., Seemayer, T. A., and Lapp, W. S. (1988). Prevention of murine graft-versus-host disease by inducing and eliminating ASGM1$^+$ cells of donor origin. *Transplantation*, **45**, 586–90.

97 Blazar, B. R., Soderling, C. C. B., Koo, G. C., and Vallera, D. A. (1988). Absence of a facilitatory role for NK 1.1-positive donor cells in engraftment across a major histocompatibility barrier in mice. *Transplantation*, **45**, 876–83.

98 Charley, M. R., Mikhael, A., Bennett, M., Gilliam, J. N., and Sontheimer, R. D. (1983). Prevention of lethal, minor-determinate graft-versus-host disease in mice by the *in vivo* administration of anti-asialo GM1. *J. Immunol.*, **131**, 2101–3.

99 Agah, R., *et al.* (1989). Potent graft antitumor effect in natural killer-resistant disseminated tumors by transplantation of interleukin 2-activated syngeneic bone marrow in mice. *Cancer Res.*, **49**, 5959–63.

100 Hoffman, R. A., Jordan, M. L., Ascher, N. L., and Simmons, R. L. (1988). The contribution of natural killer cells to the allograft response. *Transplant. Proc.*, **20**, 186–8.

101 Lord, R. H. H., Padberg, W. M., Hancock, W. W., Kupiec-Weglinski,

J. W., and Tilney N. L. (1989). Correlation of macrophage and NK cell numbers with 'activation markers' in rat cardiac allografts. *Transplant. Proc.*, **21**, 449–50.

102 Marboe, C. C., Knowles, D. M., Chess, L., Reemtsma, K., and Fenoglio, J. J. (1983). The immunologic and ultrastructural characterization of the cellular infiltrate in acute cardiac allograft rejection: prevalence of cells with the natural killer (NK) phenotype. *Clin. Immunol. Immunopathol.*, **27**, 141–51.

103 Hancock, W. W. (1984). Analysis of intragraft mechanisms associated with human allograft rejection: immunohistologic studies with monoclonal antibodies. *Immunol. Rev.*, **77**, 61–84.

104 Heidicke, C. D., *et al.* (1985). Lack of evidence for an active role for natural killer cells in acute rejection of organ allografts. *Transplantation*, **40**, 441–4.

105 Bradley, J. A., Mason, D. W., and Morris, P. J. (1985). Evidence that rat renal allografts are rejected by cytotoxic T cells and not by nonspecific effectors. *Transplantation*, **49**, 169–75.

106 Norin, A. J. and Kamholz, S. L. (1989). Evidence of intragraft interleukin-2-activated killer cells and allospecific cytolytic T lymphocytes in rejecting lung allografts. *Transplantation*, **48**, 856–62.

107 Kirby, J. A., Forsythe, J. L. R., Proud, G., and Taylor, R. M. R. (1989). Renal allograft rejection: possible involvement of lymphokine-activated killer cells. *Immunology*, **67**, 62–7.

108 Nietosvaara, Y., Renkonen, R., Mattila, P., and Hayry, P. (1990). Cytotoxic lymphocytes and lymphokine-activated killer cell-mediated lysis against rat kidney components. *Transplant. Proc.*, **22**, 127.

6 Natural killer cells in tumour biology

C. E. LEWIS and J. O'D. McGEE

1   Introduction

The identification of numerous types of infiltrating immune cells in malignant tissues (1, 2) prompted the suggestion that this phenomenon may reflect, in part, the attempt by the host's immune response to combat the presence of a neoplastic cell population. Indeed, some types of rodent malignant tissues can evoke a T-cell-mediated immune response by the expression of tumour-associated antigens. However, after an extensive search for an effective form of specific antitumour immunity, it was realized that many tumours were non-immunogenic (4). This was promptly followed by the discovery of a tumouricidal, non-T effector cell whose cytotoxic effects were not MHC-restricted or dependent on prior immunization and/or the immunogenicity of the tumour target cell. The term 'natural killer' or NK cell was then coined to describe the cell type thought to exert this form of cytotoxicity towards selected tumour cell types *in vitro* (now widely known as 'NK activity', see Chapter 1) (5, 6). Since these early studies, numerous studies have outlined the identity of the cell performing human NK activity as being predominantly $CD3^-/CD4^-/CD16^+/CD56^+$ large granular lymphocytes (7, 8). As such, the cytolytic function of NK cells can be distinguished from the natural cytotoxicity displayed by non-MHC restricted T-cells (9), B-cells (10), and macrophages (11). The contribution of these latter cell types to the antitumour defence mechanisms of the body has been extensively reviewed in recent works (12, 13) and will not be discussed further here.

A plethora of further studies has unearthed the multifaceted role of NK cells not only in tumour biology, but also in homeostatic and antimicrobial mechanisms, as well as modulation of neural and endocrine systems in the body. These wide-ranging functions are mediated, to a large extent, by the secretion of numerous cytokines by NK cells (see Table 6.1). Indeed, this wide range of activities has prompted some researchers to voice concern as to whether the tumouricidal activities of NK cells may simply be a somewhat inefficient by-product of a considerably more elaborate and efficient defence system mounted by this cell type against viral, bacterial, and parasitic invaders (3, 13).

In this chapter we review the evidence, such as it is in 1991, for the role of NK cells in the response mounted by the natural immune system to the presence of a tumour. Set, as this chapter is here, amongst a number of reviews describing the numerous and wide-ranging functions of this cell type in homeostatic and pathological conditions, it is hoped that the reader will form his/her own conclusion as to the importance of the contribution made by NK cells to the host–tumour interaction. Or, indeed, whether NK cells should be viewed as highly versatile cells which are capable of contributing various functions to a number of body processes.

Table 6.1 Cytokines released by human peripheral blood NK cells or NK cell clones: the current list

Cytokine	Type of stimuli	Ref.	Directly cytotoxic or cytostatic for selected tumour cell types	Ref.
IL-1	Bacterial products IL-2	102	Yes	124
IL-2	Lectin PMA IFNγ NK target cells	49, 102, 120 123	Yes	125
IL-4[a]	Lectin IFNγ	102, 121	NT	
TNFα	IL-2 NK target cells	102, 105	Yes	105, 126
TNFβ	NK target cells	102	Yes	102, 126
IFNα	Virus infected target cells NK target cells IL-2	102	Yes	102
IFNγ[a]	IL-1, IL-2 IL-4, IL-6 PMA, basic FGF NK target cells	49, 102, 105 139	Yes	105, 124
CSF	Lectins	102, 122	NT	
NKCF	Mitogens NK target cells	102, 123	Yes	123, 126

[a] Cytokine also secreted by human tumour-infiltrating NK cells (67).
NT, not tested.
Abbreviations: IL, interleukin; TNF, tumour necrosis factor; IFN, interferon; CSF, colony-stimulating factor; NKCF, natural killer cytolytic factor; PMA, phorbol 12-myristate 13-acetate.

2 Role of NK cells in the host–tumour interaction

The potential role of NK cells in the multifactorial innate immune mechanisms mounted by the body to the neoplastic condition can be divided into three broad categories: (1) early recognition of the onset of malignant transformation and initial development of a tumour; (2) response to the proliferation of tumour cells and increased tumour mass; and (3) attempts to regulate the metastatic spread of tumour cells in the blood and lymphatic system. Since these categories can be considered to represent progressive levels in the development of a given malignancy, it is in this order that we have chosen to discuss and elaborate each area of potential antitumour NK cell activity.

2.1   NK cells in immunosurveillance against the development of tumours

The possibility that NK cell status and function may play a part in detecting and regulating in some way the initial induction of spontaneous or carcinogen-induced tumours has been extensively investigated. However, the efficacy of NK cells in this form of immunological surveillance is debatable since, by definition, the very existence of the latter two categories of natural antitumour immunity, indicates that this first line of antitumour defence must have failed or, at least, been only partially effective. As argued by Stutman and co-workers (3, 13), escape from host surveillance may denote either the ability of malignant cells to subvert such initial defence mechanisms of the host, and/or that such cellular mechanisms do not exist as true surveillance systems. The somewhat fragmented and, at times, conflicting evidence for the involvement of NK cells in such mechanisms has mainly resulted from studies attempting to correlate the risk of tumour appearance and development with spontaneous or induced alterations in NK activity.

2.1.1   *Congenital abnormalities in NK cell function*
Blood NK cells from patients with Chediak–Higashi syndrome display markedly suppressed NK activity *in vitro*. By contrast, the antitumour activities of monocytes, T-cells, and granulocytes remains unaffected. Attempts have been made to correlate this deficiency with the higher susceptibility of these patients to the development of lymphoproliferative diseases (14). Also, normal individuals from families with a high incidence of melanoma (15) or various forms of carcinoma (16) express lower NK activity than from families without such a family history.

A low level of NK activity has also been reported in patients with X-linked lymphoproliferative syndrome; a phenomenon which has been claimed by some investigators to be linked to the increased incidence of lymphoma in this condition (17). However, a subsequent and more detailed study of the NK deficiency in such patients revealed this to be merely a consequence of the lymphoproliferation and not a causal or permissive factor (18).

Other studies have employed strains of mice with genetically determined alterations in NK activity such as those bearing the beige mutation [reviewed in (3, 19)]. Although beige mice, which have a marked, albeit incomplete, deficiency in NK activity exhibit an age-dependent tendency to develop lymphomas (20), they do not show an increased incidence of spontaneous or carcinogen-induced tumours (3, 21). Thus, such studies have yet to provide definitive evidence for a central role for NK cells in tumour surveillance since tumour development is not necessarily associated with, or preceded by, reduced NK activity.

2.1.2 *Chemically induced alterations in NK cell function: effect of immunosuppressive drugs and carcinogens*

Patients receiving kidney allotransplants and concomitant high-dose immunosuppressive therapy are more susceptible to the development of various forms of cancer (22). NK deficiency has also been induced by employing the general immunosuppressive effect of carcinogenic agents. As such, this has become one of the most popular models for testing the role of effector cells in antitumour activity, the prediction being that the depressed immune function of mice exposed to these chemicals would lead to increased tumour frequency. Indeed, numerous studies have provided data to support this prediction. Treatment of mice with the appropriate doses of such carcinogens as urethane, dimethylbenzanthracene and methylcholanthrene, induced a suppression of NK activity which appeared to correlate with a higher incidence of some tumour types [reviewed in (3, 19)]. However, it should be noted that other studies have shown that the depressive effect of urethane is dependent upon the strain of mice being used and does not have a widespread or marked correlation with the risk of tumour development (23, 24). Moreover, high doses of 7,12-dimethylbenzanthracene, whilst depressing NK activity, did not alter the ability of mice to eliminate inoculated tumour cells from their lungs (25). As mentioned earlier, and reviewed by Stutman (3), certain tumour types are seen in control and NK-deficient beige mice administered 10-dimethyl-s, 2-benzanthracene, 3-methylcholanthracene, and urethane. Indeed, a number of studies were unable to report a clear correlation between the incidence of carcinogen-induced tumours and levels of NK activity (3, 19, 23, 26). This has led some to suggest that these paradoxical findings may be due to inherent differences between various strains of mice in their resistance to the carcinogenic (that is, transforming) effects of some agents (19). If this were the case, resistant strains might not show increased tumour incidence, even in the presence of carcinogen-induced reductions in NK activity.

From the above it can be seen that experimental manipulation of NK activity has yet to provide firm evidence to support the possibility that NK cells play an important role in regulating tumour development.

2.2 NK cells in the regulation of primary tumour growth

That NK cells might have an important role in inhibiting the proliferation of tumour cells *in vivo* has been suggested by the following observations. Experimentally-induced and transplanted tumours grow more rapidly in NK-deficient animals (e.g. homozygous beige mice) [for reviews see 3, 16,

27, 28]. A second line of experimental evidence for the above has been provided in experiments indicating that this growth regulation is dependent upon
the susceptibility/resistance of tumour cells to NK cell cytoxicity (as defined in
in vitro assays) (3, 16). Thirdly, in nude mice whose NK activity had been
depressed using anti-asialo-GM_1 serum, NK susceptible, but not NK-
resistant, tumour cell lines were seen to exhibit accelerated growth *in vivo*
(29). Moreover, marked differences in the growth rate of NK-susceptible and
NK-resistant B16 melanoma cell lines transplanted into normal C57BL/6
mice were evident (27). Fourthly, in lethally-irradiated mice receiving bone
marrow transplants from strains of mice with various levels of NK activity,
recipients of 'high NK' transplants showed an increased resistance to the
growth of such NK-susceptible cell lines as YAC lymphoma cells (30).

Factors influencing the growth of tumours *in vivo*, whether cell-
mediated or humoral in nature, are likely to be part of a complex, multifactorial interaction between infiltrating immune cells and tumour cells.
The presence of immunophenotyped NK cells has been recorded, albeit in
smaller numbers than some other tumour-infiltrating cells (usually less
than 10 per cent of the total infiltrate), within such tumours as malignant
melanomas (31), squamous cell carcinomas and keratocarcinoma (the latter
type resemble squamous cell carcinoma but spontaneously regress) (32),
and malignant breast lesions (33). Indeed, in these studies, NK cells were
often seen to make intimate contact with the neoplastic cell population, as
well as simply taking up residence in the stroma of the tumour. As such,
the contribution of NK cells to tumour cell growth regulation may involve a
direct effect on tumour cells themselves via cytolysis of NK-susceptible
target cells and/or by the release of cytokines with direct cytotoxic/
cytostatic effects on tumour cells. As shown in Table 6.1, these include
interleukin-1 (IL-1), IL-2, tumour necrosis factor (TNF), natural killer
cytolytic factor (NKCF) and/or various forms of interferon (IFN). An
indirect effect of NK cell-derived cytokines on tumour cells via the modulation of the growth-regulating properties and/or cytotoxic properties of
other infiltrating cell types (e.g. macrophages, T-, and B-cells) may also be
involved.

2.3   NK cells in metastasis

Evidence for the participation of NK cells in the response of the host to a
tumour has been provided in studies suggesting their ability to eliminate
blood-borne metastasizing tumour cells and inhibit the development of
metastatic foci.

Both the survival in the bloodstream of tumour cells intravenously
inoculated into mice, and the rate of formation of metastatic deposits have
been correlated with the level of NK activity of recipient mice (19, 34, 35),
with highest levels of the former being evident in beige mice with markedly

reduced NK activity (34, 35). Additionally, the elimination of inoculated cells could be enhanced by pretreating recipient mice with agents known to augment NK activity, such as *C. parvum*, BCG, poly1:C and IFN (36–38). Suppression of NK activity by cyclophosphamide treatment, cortisol, β-oestradiol, urethane, anti-asialo-GM_1 or anti-NK serum decreased the rate of clearance of tumour cells and augmented the development of metastatic foci (39–42). These effects could be abrogated by transfusion of treated mice with spleen cells or blood NK cells from healthy donors, a process known to enhance NK cell function *in vivo* (34, 43). These studies accord well with the finding that in patients with head and neck or colorectal cancers, the appearance of metastatic deposits or the recurrence of malignancy after a disease-free interval was accelerated in individuals with low NK activity (44). Thus, in some circumstances, the level of NK activity appears to be of prognostic significance.

NK activity has been shown to be compartmentalized within the body, with variations between the blood, lymph nodes, and selected tissues examined to date. This has recently been extended by the observation that although NK activity in the peripheral blood of cancer patients is often reduced compared to that of healthy donors (44, 45), it is often elevated above that of NK cells obtained directly from the tumour site or associated lymph nodes (46, 47) (see Section 3.1). This lends credence to the recent suggestion that NK cells may be more involved in the clearance of metastasizing tumour cells from the blood than in the perivascular space or body organs (3, 19).

Although a number of investigators have been able to correlate the metastatic potential of tumour cell lines with their resistance to NK activity *in vitro*, this has been shown mainly in animals with high levels of NK activity, and is not true for mice with low NK reactivity, such as young nude mice (48). A relationship between these parameters has yet to be unequivocally demonstrated (3, 19). Various so-called NK-resistant cell lines have been detected in the blood of recipient mice and been rapidly eliminated, whilst other NK-susceptible tumour cell lines were seen to escape the antimetastatic activity of NK cells and establish tumours (3, 19). These findings could indicate that: (1) susceptibility to NK lysis *in vitro* may not equate fully with that *in vivo*; and/or that (2) the mechanism(s) which detect and eliminate metastasizing cells are complex and interactive, and only partly mediated by NK cells.

3 Factors involved in the regulation of NK cell tumouricidal activity

The accompanying chapters in this volume indicate the multifunctional nature of NK cells, inasmuch as they are able to perform not only a range of tumouricidal functions, but also diverse immuno-inflammatory activities (many of which are mediated throuh the release of various pleiotropic cytokines), and a number of antimicrobial actions. However, since the contribution made by NK cells to the latter categories of function has only recently been revealed, most studies of the regulation of NK cell function have focused on NK cell cytotoxicity towards tumour cells. As such, the following discussion is largely confined to this area of NK cell biology.

NK cells, like many other effector cells, exhibit marked functional heterogeneity depending upon their location in the body and exposure to presumptive tissue-specific microenvironmental influences (e.g. blood, spleen, lungs. liver, tumour) (8, 19, 45–47, 49–51). In addition, NK activity is known to vary with the age (50, 52), sex (53), circadian/circannual rythms (reviewed in 54), diet (54), and exposure of the donor to stress (56). Levels of NK activity can also vary markedly between different individuals and different strains of mice (although the latter has often been equated with distinct genetic differences).

Taken together, these findings indicate the tumouricidal activity of NK cells to be a complex phenomenon under the influence of a wide range of factors in the body. Recent studies have unearthed the identity of many of these regulatory factors, but relatively little about the mechanism(s) responsible for their effects. In theory, regulation of NK activity could occur at several levels in the lytic process: (1) recognition of the target cell by the NK cell (e.g. altered expression of the putative 'NK receptor' and/or NK signalling molecules on the surface of the target cell); (2) effective binding of the NK cell to the target; (3) release of pore-forming elements and/or cytolytic cytokines by the NK cell in the vicinity of the target cell; and (4) disengagement and recycling of the NK cell for repeated target lysis. As discussed more fully below, NK-regulatory factors may also influence NK activity indirectly by stimulating/inhibiting the 'suppressor' activity of other cell types.

Since the antitumour functions of NK cells have been shown to have potential in the treatment of human cancers (see Section 3.2.1), unravelling the complex network of factors which have the potential to influence NK activity, and their interactions within the body could have profound implications for the design of effective therapeutic strategies involving this cell type.

3.1 Inhibition of NK activity

Exposure of murine or human NK cells to such agents as anti-asialo-GM_1 (29), TGFβ (57) or the prostaglandins PGE_1, PGE_2, PGA_2 (58–60) *in vitro*, induces a dose-dependent suppression of NK activity. Although the mechanism(s) for this inhibition have yet to be fully elucidated, in the case of prostaglandins the effect is rapidly reversible and has been shown to act at a post-target cell-recognition locus [reviewed in (8)].

A second cellular mechanism by which NK activity can be inhibited involves the generation or activation of 'suppressor' cell types. These are thought to comprise mainly macrophages and null cells, which are capable of inhibiting the cytotoxic efficacy of NK cells *in vitro* (8). That such 'suppressor' cells play a physiological role in the regulation of NK cells *in vivo* was suggested by the finding that similar cells are present in neonatal mice, which normally have low NK activity, and in strains of mice with low levels of NK activity (61). Furthermore, tissues with low NK activity (thymus, lymph nodes, and peritoneal cavity) are rich in such 'suppressor' cell activity (8, 62, 63). Little information is currently available on the ways in which these two cell types are activated, or the mechanism by which they subvert the lytic activity of NK cells, but a role for soluble mediators released by suppressor cells has been demonstrated (8). Indeed, a recent study has shown that one such suppressive mediator released by human monocytes is PGE_2 (64). The release of reactive forms of molecular oxygen, particularly hydrogen peroxide or superoxide, by monocytes has also been suggested (8, 65).

Cells which suppress NK activity are also found in individuals bearing malignant tumours. This has been correlated with the low level of NK activity seen in these patients. It is of note that such suppressed NK activity has now been demonstrated in the blood of patients with gastrointestinal (45), head and neck (44), or cervical (66) tumours. This may result from enhanced systemic 'suppressor' cell activity and/or the release of one or a number of cytokine(s) by the tumour which directly inhibit NK cell lytic behaviour. In support of the latter possibility, Pillae and co-workers (66) recently showed the presence of NK-suppressive factors in the serum of cervical cancer patients, in which they also recorded reduced NK activity. Additionally, within the environment of tumours, where the concentration of such putative tumour-derived inhibitory signals would supposedly be highest, NK activity is markedly reduced compared to that in the peripheral blood of the same patients (46, 47). However, this does not exclude the possibility that 'suppressor' cells may be present within the tumour and be a source of such NK-suppressive cytokines. Indeed, various studies have indicated the presence of large numbers of infiltating macrophages in solid tumours which could act in this way (1, 2). A role for intratumoural 'suppressor' macrophages was supported by data from co-culture experiments indicating the NK activity of normal human cells to be inhibited by tumour-derived macrophages. Our recent work has provided evidence to suggest that the neoplastic cell population itself may also be involved in the regulation of such effector cells at the tumour site. Secretion of IFNγ and IL-4 by tumour-infiltrating NK cells is enhanced by factor(s) released by tumour cell lines (49, 67). Pilot studies have implicated epidermal growth factor secretion by tumour cells in this phenomenon (C. O'Sullivan, C. E. Lewis and J. O'D. McGee, unpublished observations). Thus, it is possible that tumour cells could play a part in the regulation of local, intratumoural NK activity, both directly by the action of tumour cell cytokines on NK cells, and/or indirectly via their effect on 'suppressor' cell activity.

Many experiments using animal models have demonstrated that both the success of primary tumour implantation and growth, and the rate, size, and incidence of subsequent metastases increase after surgical manipulation and stress (56, 68, 69). Furthermore, a marked impairment of NK activity occurs in patients with ovarian (70), head and neck (71), and breast (72) cancer following surgery. Although this phenomenon has also been seen after non-malignant surgery (73), our recent study indicates that other functions of NK cells may not be similarly suppressed by surgery. Cytokine secretion of human NK cells was augmented, rather than inhibited, following major abdominal surgery (74). The cellular mechanisms subserving this effect have yet to be fully outlined, but appear to involve a number of immune dysfunctions, of which the generation of various suppressor cell types and impairment of NK cell activity are but two. Surgery-related suppression of NK activity appears to require cell-to-cell contact and to occur in multiple body compartments (56). However, most authors agree that surgical stress in mice alters NK cell cytotoxicity by generating splenic cells which effectively compete with NK cells for tumour target binding sites, and by inducing an acute decrease in maximal NK cell recycling capacity during the lytic process (75). Studies are currently under way to determine the mechanism of human peri- and post-operative impairment of NK cytotoxic activity. It is hoped that this will lead to the development of new immunotherapeutic strategies for patients undergoing solid tumour resection.

Over the past few years, it has become apparent that emotional, as well as physical, stress can result in marked alterations in the cytotoxic efficacy of human NK cells towards tumour targets. Individuals experiencing depression (76), schizophrenia (77) or bereavement (77), exhibit profoundly suppressed NK activity. Furthermore, in breast cancer patients, NK activity and survival rate could be correlated with emotional status and perceived social support (78). Such studies not only indicate the functional interaction which occurs between the immune and neuro endocrine systems of the body, but also highlight the potential contribution of stress-related events to the efficacy of such effector cell populations as NK cells in the treatment of human malignancy.

Since the effects of various forms of stress on the immune system [reviewed in (79)] appear to be mediated by the release of 'stress-related' hormones, it is entirely feasible that the effects of physical and emotional stress on NK activity may be mediated, in part, by such endocrine factors. Indeed, a wide range of hormones and neurotransmitters which alter natural effector cell functions have recently been identified (80). The effects of such factors on NK activity are listed in Table 6.2. In the case of

Table 6.2   Endocrine factors involved in the regulation of NK activity

Factor	Effect on NK activity
Pituitary hormones	
GH	Stimules[a]
PRL	Inhibits[b]
LH	Inhibits
Steroids	
Oestrogen	Inhibits/No effect[a,b]
Glucocorticoids	Inhibits[c]
Progesterone	No effect
Testosterone	No effect
Opioids	
α-Endorphin	No effect
β-Endorphin	Stimulates[c]
Met-Enkephalin	Stimulates
Morphine	No effect
Others	
VIP	Stimulates/Inhibits[b]
Serotonin	Stimulates
Melatonin	Stimulates/Inhibits[a,b]
Prostaglandins A and E	Inhibits
Adrenaline	Stimulates/Inhibits[b,c]
Thyroxine	Stimulates/Inhibits[a,b]

[a] Response only demonstrated *in vivo* (i.e. not necessarily a direct effect on NK cells).
[b] Response depends on duration of exposure or administration.
[c] Receptors present on NK cells, as demonstrated by binding studies or competitive antagonism.
[Adapted from (80).]
GH, growth hormone; PRL, prolactin; LH, luteinizing hormone; VIP, vasoactive intestinal peptide.

stress-related changes in NK activity, the finding that prolactin and cortisol can both exert an inhibitory influence on NK cell cytotoxicity is particularly relevant (80). Although patients with hyperprolactinemia have suppressed NK activity (81), this effect is thought to be via the action of prolactin on NK cell progenitors and/or by inducing hormonal/metabolic changes which then act on mature NK cells (82). Prolactin does not appear to have any direct modulatory affect *per se* on NK cells *in vitro* (80, 82). Exposure of NK cells to glucocorticoids *in vivo* or *in vitro* reduces NK activity (80, 83, 84). Recent evidence indicates that brain corticotrophin-releasing factor (CRF) plays a physiological role in co-ordinating stress-induced suppression of NK activity in mice via its effects on autonomic and neuroendocrine activity (69). Of relevance to this finding are data indicating that adrenaline (which can be stimulated by CRF) inhibits NK activity via a rapid action if actually added during the cytotoxicity assay (although preincubation with this amine enhances NK activity) (85). Although a growing body of research data supports the view that a number of neurohormones/transmitters, such as serotonin, neurotensin, substance P, and vasoactive intestinal peptide (VIP) can influence various effector cells, their effect on NK activity has yet to be clearly established (80) (Table 6.2).

3.2   Stimulation of NK activity

Murine and rat studies have shown that inoculation with NK-susceptible tumour cells, certain viruses or immune adjuvants, such as BCG, *C. parvum* of poly1:C can augment NK activity. That this is, at least in part, a direct effect on NK cells has been confirmed in subsequent *in vitro* studies. Some evidence suggests that the effects of these agents may be partially mediated by boosting IFN release, which then serves to activate NK cell cytotoxicity (see below).

Recently, the antitumour potential of flavonoids (particularly flavone acetic acid, FAA) has come to light. FAA can be directly cytotoxic for tumour cells *in vitro*, although it should be noted that this effect is only seen at high concentrations and after long periods of exposure (86, 87). Since this compound is significantly more tumouricidal against the same tumour cell types *in vivo*, suggestion was made that FAA may also act to augment the antitumour actions of lymphocytes [reviewed in (86)]. Indeed, the cytotoxic activity of NK cells can be directly stimulated by FAA exposure (86–88). Various murine studies have indicated that FAA may be most therapeutic when combined with other biological response modifiers (e.g. IL-2) (89, 90), and that the beneficial effect of FAA is abrogated by suppression of FAA-induction of NK activity in tumour-bearing mice (90). In these studies, FAA-induced augmentation of NK activity could be correlated both with increased levels of IFNα/β and induction of long-lasting tumour immunity. Trials are currently underway to examine the activity of this compound when administered to cancer patients in combination with other NK-boosting agents. However, a recent clinical trial involving treatment of patients bearing metastatic melanomas with FAA and/or IL-2 failed to show any additional effect of FAA to that exerted by IL-2 alone (91). Moreover, a study of the action of FAA treatment in colon cancer has shown that the effect of FAA on NK cytotoxicity may be compartmentalized within the body, with NK activity stimulated by FAA in the spleen, but depressed at the tumour site (92). Indeed, we have recently demonstrated that FAA inhibits, rather than augments, IFN-γ release by human NK cells *in vitro* (D. Cane, C. E. Lewis, and J. O'D. McGee; unpublished observations).

Table 6.3 Cytokine regulation of NK and LAK cell activity

Activity	Cytokine	Effect	Ref.
NK	IFNα	Stimulates	127
	IFNβ	Stimulates	128
	IFNγ	Stimulates	129
	IL-2	Stimulates[a]	130
	IL-3	Stimulates	131
	IL-4	No effect	132
	IL-6	Stimulates[a]	133
	IL-8	ND[b]	
	TNFα	Stimulates	134
	TGFβ	Inhibits	57
	FGF	ND[a]	
	PGE$_2$	Inhibits	135
LAK	IFNα	Inhibits	136
	IFNβ	Stimulates/Inhibits[c]	136
	IFNγ	Stimulates/No effect	136, 137
	IL-1	Stimulates/No efect	138
	IL-2	Stimulates	138
	IL-3	Inhibits	139
	IL-4	Stimulates/Inhibits[c]	114, 132, 138
	IL-6	No effect	138
	GMCSF	Stimulates/No effect[c]	138
	PGE$_2$	Inhibits	58

ND, not determined.
[a] Cytokine which augments the spontaneous release of IFNγ by human NK cells *in vivo* (49).
[b] IL-8 has recently been shown to inhibit the spontaneous release of IFNγ by human NK cells *in vitro* (49).
[c] Effect dependent on whether LAK precursor cells are exposed to this cytokine before, during or after IL-2 stimulation (114).

Platelet aggregation and fibrin coagulation on the surface membrane of tumour cells has been suggested as a possible mechanism by which tumour cells are protected from destruction by NK cells. Indeed, platelet aggregating and procoagulant activity of tumour cells has been repeatedly demonstrated (93, 94). It is, therefore, not surprising that exposure to anticoagulant drugs, such as heparin and warfarin, has been shown in mice to increase the efficacy of NK cell cytotoxicity, rate of elimination of some forms of inoculated tumour cells from the blood, and to decrease metastasis formation [reviewed in (19)].

As discussed in Chapter 1 (Section 5.3.2) and illustrated in Table 6.3, a number of cytokines exert regulatory effects on NK activity. Interferons, α, β, and γ, enhance NK activity, although their effects on this type are complex and operate at various levels of the lytic process. Some investigators have claimed that IFNs can increase the ability of NK cells to recognize and bind certain tumour targets, although this does not appear to be the case for all NK-susceptible tumour cells. More convincing evidence has been presented indicating that IFN boosts the rate of NK lysis of tumour cells by increasing the degree of recycling of NK cells, thereby increasing interactions between NK cells and their targets and facilitating the lysis of multiple target cells during the cytotoxicity assay (8, 95). However, an interesting caveat to these direct effects of IFN on NK cells, is that it appears that IFNγ may also be able to protect tumour cells from NK cell lysis, possibly by a mechanism involving alterations in the expression of MHC Class I (96). As NK cells are spontaneously lytic towards tumour cells lacking MHC gene products, some investigators have suggested that the reduced or absent expression of these molecules on tumour cells may play a part in the mechanism by which NK cells recognize and lyse targets. For this reason it was proposed that IFNγ may induce resistance to NK killing by enhancing MHC Class I expression (96). However, recent reports have been unable to directly correlate alterations in MHC Class I expression with NK lysis of tumour targets (97).

The search for endogenous neural and endocrine immunomodulators in recent years has revealed a number of factors with NK-boosting actions. Amongst the most prominent of these are the opioids, met-enkephalin (98) and β-endorphin (99), which can exert a direct and rapid effect on NK cells *in vitro*. Faith and co-workers (98) suggest that levels of certain opioids may actually fall with the onset of certain diseases or chronic stress and that replacement therapy in cancer patients may provide a new and useful form of immunotherapy. This has been given credence by their finding that the cytotoxic activity of NK cells isolated from patients with various forms of advanced cancer could be markedly enhanced by exposure to such opioids *in vitro* (98). Moreover, we have recently demonstrated a marked stimulatory effect of the opioids, β-endorphin, and met-enkephalin, on the cytokine-secreting activity of human NK cells *in vitro* (D. Carroll, C. E. Lewis, and J. O'D. McGee, unpublished observations).

Various interleukins are known to stimulate the cytotoxic (Table 6.3) and/or IFN-secreting activity (49) of NK cells. However, by far the most potent enhancer of NK activity discovered to date is interleukin-2 (IL-2).

3.2.1   IL-2 and the lymphokine-activated killer (LAK) cell phenomenon

First described as a T-cell growth factor in 1975, interleukin-2 (IL-2) is now known to stimulate a number of other cell types including the proliferation, cytokine production and natural cytotoxicity of human NK cells [extensively reviewed in (8, 19, 49, 100)] (see also Chapter 1). As with the influence of IL-2 on T-cells, the relatively slow action of this cytokine on NK proliferation is dependent upon interaction with one form of IL-2 receptor, TAC.

However, unlike T-cells, the expression of this reeptor on NK cells does not require prior stimulation by mitogens or antigen. Indeed, the effect of IL-2 on the other two categories of NK cell function listed above can be induced rapidly prior to the expression of TAC (which only usually appears on NK cells within the first 2 days of IL-2 exposure *in vitro*). In fact, IL-2-induced augmentation of NK cell cytotoxicity will occur even in the presence of anti-TAC antibodies (8). Taken together, these findings indicate that stimulation by IL-2 of the lytic capacity of NK cells is exerted via non-TAC receptors (see Chapter 1 for the biochemical events triggered by IL-2 activation of NK cells).

The observation that a killer cell with significantly greater cytotoxicity *in vitro* and *in vivo* could be generated after short-term culture of peripheral blood lymphocytes with IL-2, has now become something of a milestone in tumour immunology (8, 100). These lymphokine-activated killer (LAK) cells, as they became known, share many properties with NK cells, but will also lyse a variety of fresh autologous, allogeneic, and xenogeneic tumour targets. Although the progenitor cells in human blood or splenic lymphocytes which give rise to LAK activity in the presence of IL-2 have been shown to be $CD3^-/CD16^+/CD56^+$ NK cells, many NK-resistant cell lines have been shown to be susceptible to LAK activity *in vitro*. Some minor LAK activity is seen to be mediated by a small subpopulation of $CD3^+/CD4^-/CD8^-$ T-cells which also express CD56 (8, 100, 101). These data indicate that LAK activity should be regarded as a category of function, rather than be designated to a new or distinct effector cell population. Alternatively, it is possible that we still have much to learn about the differentiation pathway(s) of NK/LAK cells.

The cellular mechanism(s) by which LAK cells lyse their tumour targets is unknown at this time, although it is widely believed to be similar to that employed by NK cells. This involves the release of such pore-forming substances as serine-esterase and perforins, and/or the release of the cytokines, IFNγ and TNFα, which have been shown to exert a cytostatic and/or cytotoxic effects on certain tumour cells *in vitro* (8, 102–105) (Table 6.1). Confirmation of this prediction was recently provided in the form of data showing the secretion of elevated levels of IFNγ and TNFα by human LAK cells stimulated by contact with tumour targets. These factors were seen to mediate the slow-acting cytolytic action of LAK cells towards tumour cell lines *in vitro* (105). However, a note of caution concerning the potential benefits to the host of the presence of these two cytokines at the interface between the immune system and the tumour should be mentioned. Recent studies have shown that IFNγ increases the metastatic potential of certain tumour cell types (106, 107), the mechanism of which appears to involve protection of targets from NK cell-mediated lysis (see above) (96, 97, 107). Another action of IFNγ is to stimulate the release of TNFα by monocytes/macrophages. TNFα may have tumour-promoting properties since it is reported to stimulate angiogenesis (108), promote tumour invasion in ovarian cancer xenograft models (109), and enhance the metastic potential of some tumour cell types (110). Hence, the fact that both LAK cells and macrophages have the potential to be stimulated at the tumour site to secrete elevated levels of this cytokine may not be entirely advantageous to the host. However, such considerations of the advantages of products secreted by LAK cells may prove only to be relevant to their systemic activity, since these cells, when generated *in vitro* and injected into the bloodstream of the host, fail to show a preferential 'homing' mechanism for the tumour and appear only in low numbers at this site (111).

The fact that cancer patients do not appear to have high endogenous levels of LAK activity prior to stimulation with exogenous IL-2, may be due to the presence of suppressive factors in the serum (and possibly at the tumour site). Indeed, serum derived from such individuals can exert a marked inhibitory effect on IL-2 induction of LAK activity from blood lymphocytes of healthy donors (112). The identity and source of this/these soluble mediator(s) has yet to be ascertained, but a recent study by Guillou and his co-workers has shown that tumour cells themselves can secrete factors which inhibit LAK induction by a direct action on LAK cell progenitors, via a mechanism which does not require the presence of macrophages or macrophage-derived products (113).

A number of other immunomodulatory cytokines are also known to influence LAK activity (Table 6.3). The most prominent and well researched of these include the stimulatory effect of IFNγ and the dual effects of IL-4. Exposure of LAK progenitors to IL-4 markedly inhibited the degree of LAK activity induced by subsequent incubation with IL-2. Alternatively, once the cells had been sufficiently pre-activated by IL-2, IL-4 was seen to stimulate, rather than suppress, LAK activity (114).

Emphasis has been placed on the fact that LAK cells are able to lyse some forms of freshly isolated tumour cells. This is particularly marked in studies using fresh human leukaemia cells [reviewed in (19, 100)]. However, in a recent review, Herbermann points out that unstimulated NK cells also show significant cytotoxic activity against many fresh solid tumour or leukaemic cells (8). These data, generated largely *in vitro*, conflict somewhat with therapeutic trials which have attempted to utilize the antitumour functions of unstimulated or stimulated (but non-LAK) NK cells. It is interesting to note that such NK stimulants as interferon (IFN), or IFN inducers, when administered to cancer patients, initially boost NK activity, but then become unresponsive to subsequent doses of the cytokine. Indeed, repeated exposure to IFN elicited a suppression of NK activity (19). Various studies have suggested that this effect may have been mediated by IFN-induction of NK-suppressor macrophages. This possibility highlights the problem of administering multifunctional cytokines as part of a therapeutic regimen; the cascade of events triggered in the body may include antagonistic functions, and possibly side-effects which are deleterious to the recipient (see below).

In the early 1980s, patients with advanced malignancies were treated with IL-2 in the hope that endogenous NK activity would be boosted and LAK activity generated *in vivo* (8, 19, 100). Although there were various toxic side-effects of IL-2 treatment including fever, hepatic dysfunction, chills, fluid retention, weight gain, and malaise, few, if any, side-effects remained once treatment was completed. However, only minor anti-tumour effects of this form of therapy were evident. Some studies have since shown that high doses of this cytokine alone can induce tumour regression in some instances (114). A surprising finding of these early studies was that LAK cell progenitors were depleted after IL-2 treatment (19). The next logical step, therefore, was to remove the blood lympho-cytes from cancer patients, stimulate them with IL-2 to express LAK activity *in vitro*, and attempt to reinfuse them back into the bloodstream of the same patients (along with IL-2 to sustain the level of augmented killer activity). When this dual exposure of advanced cancer patients with IL-2 and LAK cells was achieved by Rosenberg's group, a number of complete or partial responses were reported, mainly with patients with melanoma and renal carcinoma, although this course of treatment was again not free of unpleasant, and in a few cases, fatal side-effects (115). Since then, a number of groups have reported varying levels of success with slight modifications of the dose and duration of the IL-2 infusion regimen (100).

The results of these clinical trials using IL-2 and LAK cells in the treat-ment of advanced malignancies are encouraging but indicate that such forms of immunotherapy are still very much at the experimental stage. The reasons for most solid tumours being relatively unresponsive have yet to be outlined and toxicity remains a problem and limiting factor.

Hopeful prospects for the future include the use of a subpopulation of IL-2-stimulated LAK cells which have been shown to adhere to plastic surfaces (called A-LAK cells). After expansion in IL-2 for 3 to 4 days, A-LAK cells generated 20–50 times more LAK activity than non-adherent cultures of LAK cells (19, 115). A study of the phenotype of A-LAK cells generated from blood lymphocytes of liver cancer patients recently indi-cated the majority to be $CD3^-/CD56^+$ cells (116). Both rat and human A-LAK cells display enhanced cytolytic activity against tumour cells from a number of solid and ascites tumours, and a broad range of cell lines, both *in vitro* and *in vivo* [most recently reviewed in (115)]. A second prospect involves the isolation and IL-2 expansion *in vitro* of tumour-infiltrating lymphocytes (TILs) (114). The cytotoxic effector cells generated in this way have been found to return specifically to the tumour site in greater numbers than LAK cells when reintroduced into the host, and to be 50–100 times more effective at tumour cell lysis (100, 115). It should be noted that this potent antitumour effect has recently been shown to be mediated mainly by CD3$^+$ T-cells, rather than NK cells (117, 118). Although TIL therapy has proved successful in human cancer studies, side-effects are still induced because IL-2 infusion is required to maintain the activated tumouricidal state of the adoptively transferred cells. However, these are reduced relative to those seen with IL-2/LAK therapy since lower doses of IL-2 are required.

Other current studies in this area involves investigation of alternative modes of delivering cytokines to the host–tumour interface which minimize systemic toxicity. One innovative approach appears to be the possibility of inserting genes, such as those for cytokines shown to synergize with IL-2 in LAK-generating systems *in vitro*, or indeed, IL-2 itself, into TILs, which it is hoped would enhance the therapeutic efficiency of such effector cells. Alternatively, a suggestion made recently involved cytokine gene manipulation of the neoplastic cell population itself, in the hope that this would ensure cytokine release at the most appropriate site in the body. Early studies using transplantable animal tumour models have shown that genetically engineered tumour cell expression of IL-2, IFNγ, or IL-4 can result in augmentation of the anti-tumour response of the recipient, followed by inhibition of tumour growth (119).

4   Summary

Taken together these studies indicate the potential of adoptive therapy using such effector cell types as the NK cell, or one of its derivatives (the LAK and A-LAK cell) in the treatment of human malignancy. However, it should be noted that the application of these forms of cancer treatment, whilst encouraging, is still in its infancy.

The many, diverse factors which can influence the natural cytotoxicity of NK cells are being reported with great frequency. Since it is likely that the number, type, and combination of such immunomodulatory influences will vary markedly between individuals, the host–tumour interaction in each patient has to be considered as unique. It has also become increasingly apparent that the antitumour potency of NK cells even differs between different compartments within the body. Moreover, we know relatively little of the contribution made by NK cell-derived cytokines to the antitumour activities of this, and other immune cell types present at the host–tumour interface. Such considerations may have to be taken into account and incorporated into the development of new therapeutic strategies for the early promise of cancer immunotherapy to be fulfilled.

Acknowledgements

Work in this laboratory has been funded by grants from the Cancer Research Campaign, UK to C.E.L. and J.O'D.M. C.E.L. also holds a Research Fellowship at Green College, Oxford, funded by Yamanouchi Inc. The careful and speedy preparation of this typescript by Miss Lesley Watts is also gratefully acknowledged.

References

1 Hamlin, I. M. E. (1968). Possible host resistance carcinoma of the breast, a histological study. *Br. J. Cancer.*, **2**, 383–401.

2 Viav, J., Bustamante, R., and Thivolet, J. (1977). Characterization of monuclear cells in the inflammatory infiltrates of cutaneous tumours. *Br. J. Dermatol.*, **97**, 1–10.

3 Stutman, O. (1989). Cancer. In *Natural immunity* (ed. D. S. Nelson), pp. 749–94. Academic Press, London.

4 Weiss, D. (1977). The questionable immunogenicity of certain neoplasms. *Cancer Immunol. Immunother.*, **2**, 11–16.

5 Rosenberg, E. B., *et al.* (1972). Lymphocyte cytotoxicity reactions to leukemia-associated antigens in identical twins. *Int. J. Cancer*, **9**, 648–8.

6 Herberman, R. B. (ed.) (1982). *NK cells and other natural effector cells.* Academic Press, New York.

7 Lanier, L. L., Phillips, A. M., Hackett, J. H., Tatt, M., and Kumar, V. (1986). Natural killer cells; definition of a cell type rather than a function. *J. Immunol.*, **137**, 2735–44.

8 Herbermann, R. B. (1989). Natural killer cells. In *Natural immunity* (ed. D. S. Nelson), pp. 71–122. Academic Press, London.

9 Lanier, L. L. and Phillips, J. H. (1986). Evidence for three types of human cytotoxic lymphocyte. *Immunol. Today*, **7**, 132–4.

10 Bykowski, M. J. and Stutman, O. (1986). The cells responsible for murine natural cytotoxic (NC) activity: a multilineage system. *J. Immunol.*, **137**, 1120–6.

11 Alexander, P. (1976). The functions of the macrophage in malignant disease. *Ann. Rev. Med.*, **27**, 207–45.

12 Patek, P. Q. and Collins, J. L. (1988). Tumour surveillance revisited: natural cytotoxic activity deters tumorigenicity. *Cell. Immunol.*, **116**, 240–9.

13 Stutman, O. (1975). Immunodepression and malignancy. *Adv. Cancer Res.*, **22**, 261–422.

14 Roder, J. C., *et al.* (1980). A new immunodeficiency disorder in humans involving NK cells. *Nature*, **284**, 553–5.

15 Hersey, P., Edwards, A., Honeyman, M., and McCarthy, W. H. (1979). Low natural killer activity in familial melanoma patients and their relatives. *Br. J. Cancer*, **40**, 113–19.

16 Strayer, D. R., Carter, W. A., Mayberry, S. D., Pequignot, E., and

Brodsky, I. (1984). Low natural cytotoxicity of peripheral blood mononuclear cells in individuals with high familial incidences of cancer. *Cancer Res.,* **44,** 370–6.

17 Sullivan, J. L., Byron, K. S., Brewster, F. E., and Purtilo, D. T. (1980). Deficient natural killer activity in X-linked lymphoproliferative syndrome. *Science,* **210,** 543–5.

18 Seeley, J. K., Bechtold, T., Purtillo, D. T., and Lindsten, T. (1982). *NK cells and other natural effector cells* (ed. R. B. Herberman), pp. 1211–18. Academic Press, New York.

19 Herberman, R. B. and Gorelik, E. (1989). Role of the natural immune system in the control of primary tumours and metastasis. In *Functions of the natural immune system* (ed. C. W. Reynolds and R. H. Wiltrout), pp. 3–37. Plenum, New York.

20 Haliotis, T., Roder, J., and Dexter, D. (1982). Evidence for *in vivo* NK reactivity against primary tumours. In *NK cells and other natural effector cells* (ed. R. B. Herberman), pp. 1399–1404. Academic Press, New York.

21 Haliotis, T., Ball, J. J., Dexter, D., and Roder, J. (1985). Spontaneous and induced primary oncogenesis in natural killer (NK)-cell deficient beige mutant mice. *Int. J. Cancer,* **35,** 505–11.

22 Lipinski, M., Tursz, T., Kries, H., Finale, Y., and Amial, J. L. (1980). Dissociation of natural killer activity and antibody-dependent cell-mediated cytotoxicity in kidney allograft recipients receiving high-dose immunosuppressive therapy. *Transplantation,* **29,** 214–19.

23 Gorelik, E. and Herberman, R. B. (1981). Inhibition of the activity of mouse NK cells by urethane. *J. Natl. Cancer Inst.,* **66,** 543–7.

24 Gorelik, E. and Herberman, R. B. (1982). Role of natural cell-mediated immunity in urethane-induced lung carcinogenesis. In *NK cells and other natural effector cells* (ed. R. B. Herberman), pp. 1415–21. Academic Press, New York.

25 Dean, J. H. *et al.* (1986). Immunosuppression following 7,12-dimethylbenzanthracene exposure in B6C3F1 mice. Altered cell-mediated immunity and tumor resistance. *Int. J. Immunopharmacol.,* **8,** 189–196.

26 Stutman, O. (1983). The immunological surveillance hypothesis. In *Basic and clinical tumour immunology* (ed. R. B. Herberman), pp. 155–94. Nijhoff, Boston.

27 Talmadge, J. E., Meyers, K. M., Prieur, D. J., and Starkey, J. R. (1980). Role of NK cells in tumour growth and metastasis in beige mice. *Nature,* **284,** 622–24.

28 Karre, K., Klein, G. O., Kiessling, R., Klein, G., and Roder, J. C. (1980). Low natural *in vivo* resistance to syngeneic leukemias in natural killer-deficient mice. *Nature,* **284,** 624–5.

29 Habu, S., *et al.* (1981). *In vivo* effects of anti-asialo GM_1. I. Reduc-

tion of NK activity enhancement of transplanted tumour growth in nude mice. *J. Immunol.,* **127,** 34–8.

30 Haller, O., Kiessling, R., Orn, A., Karre, K., Nilsson, K., and Wigzell, H. (1977). Natural cytotoxicity to human leukemia mediated by mouse non-T cells. *Int. J. Cancer,* **20,** 93–8.

31 Kernohan, N. M., Sewell, H. F., and Walker, F. (1990). Natural killer cells in cutaneous malignant melanoma. *J. Pathol.,* **161,** 35–40.

32 Markey, A. C. and Macdonald, D. M. (1989). Identification of CD16/ NKH1$^+$ natural killer cells and their relevance to cutaneous tumour immunity. *Br. J. Dermatol.,* **121,** 563–70.

33 Teisa, A. N., Sood, U., Pietruk, T., Cummings, G., Hasimoto, K., and Crissman, J. (1987). *In situ* quantification of inflammatory monuclear cells in ductal infiltrating breast carcinoma. *Am. J. Path.* **128,** 52–6.

34 Gorelik, E., Wiltrout, R., Okumura, K., Habu, S., and Herberman, R. B. (1982). Role of NK cells in the control of metastatic spread and growth of tumour cells in mice. *Int. J. Cancer,* **30,** 107–14.

35 Hanna, N. and Fidler, I. J. (1981). Expression of metastatic potential of allogeneic circulating and xenogeneic neoplasms in young nude mice. *Cancer Res.,* **14,** 438–42.

36 Djeu, J. Y., Heinbaugh, J. A., Holden, H. T., and Herberman, R. B. (1979). Role of macrophages in the augmentation of natural killer activity by polyI:C and interferon. *J. Immunol.,* **122,** 182–7.

37 Herberman, R. B., *et al* (1979). Natural killer cells: characteristics and regulation of activity. *Immunol. Rev.,* **44,** 43–70.

38 Herberman, R. B., *et al* (1980). Role of interferon in regulation of cytotoxicity of natural killer cells and macrophages. *Ann. N.Y. Acad. Sci.,* **250,** 63.

39 Hanna, N. and Fidler, I. J. (1981). The role of natural killer cells in the destruction of tumor emboli. *J. Natl. Cancer Inst.,* **65,** 801–10.

40 Gorelik, E. and Herberman, R. B. (1981). Inhibition of the activity of mouse NK cells by urethane. *J. Natl. Cancer Inst.,* **66,** 543–7.

41 Seaman, W., Blackman, M., Gindhart, T., Roubina, J., Loeb, J., and Talal, N. (1978). β-Oestrodial reduced natural killer cells in mice. *J. Immunol.,* **12,** 2193–6.

42 DeBrabancer, M., Aerts, F., and Borgers, M. (1974). The influence of a glucocorticoids on the lodgement and development in the lungs of intravenously injected tumour cells. *Eur. J. Cancer.,* **10,** 755–60.

43 Riccardi, C., Barlozzari, T., Santoni, A., Herberman, R. B., and Cesarini, C. (1981). Transfer, to cyclophosphamide-treated mice of natural killer cells and *in vivo* natural reactivity against tumours. *J. Immunol.,* **126,** 1284–8.

44 Schantz, S. P., Brown, B. W., Lira, E., Taylor, D. L., and Beddingfield, N. (1987). Evidence for the role of natural immunity in the

control of metastatic spread of head and neck cancer. *Cancer Immunol. Immunother, 25,* 141–8.

45 Monson, J. R., Ramsden, C. W., Giles, G. R., Brennan, T. G., and Guillou, P. J. (1987). Lymphokine activated killer cells in patients with gastrointestinal cancer. *Gut., 28,* 1420–5.

46 Moy, P. M., Holmes, C., and Golub, S. H. (1985). Depression of natural killer cytotoxic activity in lymphocytes infiltrating human pulmonary tumours. *Cancer Res., 45,* 57–60.

47 Tsujihashi, H., Matsuda, H., Uejima, S., Akiyama, T., and Kurita, T. (1988). Immunocompetence of tissue-infiltrating lymphocytes in bladder tumours. *J. Urol., 140,* 890–894.

48 Hanna, N. and Fidler, I. J. (1981). Relationship between metastatic potential and resistant natural killer cell-mediated cytotoxicity in three murine tumour systems. *J. Natl. Cancer Inst., 66,* 1183–6.

49 Lewis, C. E., McCracken, D., Ling, R., Richards, P. S., McCarthy, S. P., and McGee, J. O'D. (1991). Cytokine release by single, immunophenotyped human cells: use of the reverse hemolytic plaque assay. *Immunol. Rev, 119,* 23–39.

50 Lanza, E. and Djeu, J. (1982). Age-dependent natural killer cell activity in murine peripheral blood. In *Natural killer cells and other natural effector cells* (ed. R. B. Herberman), pp. 335–40. Academic Press, New York.

51 Wiltrout, R., *et al.* (1985). Role of organ-associated NK cells in decreased formation of experimental metastasis in lung and liver. *J. Immunol., 134,* 4267–72.

52 Krishnaraj, R. and Blandford, G. (1988). Age-associated alterations in human natural killer cells. *Cell. Immunol., 114,* 137–48.

53 Tilden, A. B., Grossi, C. E., Itoh, K., Cloud, G. A., Dougherty, P. A., and Balch, C. M. (1986). Subpopulation analysis of human granular lymphocytes: associations with age, gender and cytotoxic activity. *Nat. Immun. Cell Growth Regul., 5,* 90–9.

54 Pati, A. K., Florentin, I., Chung, V., Sousa, M., Levi, F., and Mathe, G. (1987). Circannual rhythm in natural killer activity and mitigen responsiveness of murine splenocytes. *Cell. Immunol., 108,* 227–34.

55 Herbert, J. R., Barone, J., Reddy, M. M., and Baclund, J. Y. (1990). Natural killer cell activity in a longitudinal dietary fat intervention trial. *Clin. Immunol. Immunopathol., 54,* 103–16.

56 Pollock, R. E. and Lotzova, E. (1987). Surgical-stress-related suppression of natural killer cell activity: a possible role in tumour metastasis. *Nat. Immun. Cell Growth Regul., 6,* 269–78.

57 Rook, A. H., *et al.* (1986). Effects of transforming growth factor β on the functions of natural killer cells: depressed cytolytic activity and blunting of interferon responsiveness. *J. Immunol., 136,* 3916–20.

58 Leung, K. H. (1989). Inhibition of human NK cell and LAK cell

cytotoxicity and differentiation by PGE$_2$. *Cell. Immunol.*, **123**, 384–95.

59 Kendall, R. A. and Targan, S. (1980). The dual effects of prostaglandin E2 (PGE2) and ethanol on the natural killer cytotoxic process: effector activation and NK cell-target cell conjugate lytic inhibition. *J. Immunol.*, **125**, 2770–7.

60 Koren, H. S., Anderson, S. J., Fischer, D. G., Copeland, C. S., and Jensen, P. J. (1981). Regulation of human natural killing. The role of monocytes, interferon and prostaglandins. *J. Immunol.*, **127**, 2007–13.

61 Cudkowitz, G. and Hochman, P. S. (1979). Do natural killer cells engage in regulated reactions against self to ensure homeostasis? *Immunol. Rev.*, **44**, 13–42.

62 Nair, M. P. N., Schwartz, S. A., Fernandes, G., Pahwa, R., Ikehara, S., and Good, R. A. (1981). Suppression of natural killer (NK) cell activity of spleen cells by thymocytes. *Cell. Immunol.*, **58**, 9–18.

63 Pollack, S. B. and Emmons, S. L. (1981). In *Mediation of cellular immunity in cancer by immune modifiers* (ed. M. A. Chigios, M. Mitchell, M. J. Mastrangelo and M. Krim). Raven Press, New York.

64 Bloom, E. T. and Babbitt, J. T. (1990). Prostaglandin E2, monocyte adherence and interleukin-1 in the regulation of human natural killer cell activity by monocytes. *Nat. Immun. Cell Growth Regul.*, **9**, 36–48.

65 Seaman, W. E., Gindhart, T. D., Blackman, M. A., Dalal, B., Talel, N., and Werb, Z. (1982). Suppression of natural killing *in vitro* by monocytes and polymorphonuclear leukocytes. *J. Clin. Invest.*, **69**, 876–88.

66 Pillai, M. R., Balaram, P., Abraham, T., Padmananhan, T. K., and Nair, M. K. (1988). Natural cytotoxicity and serum blocking in malignant cervical neoplasia. *Am. J. Reprod. Immunol. Microbiol.*, **16**, 159–62.

67 Lorenzen, J., Lewis, C. E., McCracken, D., Horak, E., Greenall, M., and McGee, J. O'D. (1991). Human tumour-associated NK cells secrete increased amounts of interferon and interleukin-4. *Br. J. Cancer*, **64**, 457–62.

68 Pollock, R. E., Babcock, G., Romsdahl, M. M., and Nishioka, K. (1984). Surgical stress suppression of murine natural killer cell cytotoxicity. *Cancer Res.*, **44**, 3888–92.

69 Irwin, M., Vale, W. and Rivier, C. (1990). Central corticotropin-releasing factor mediates the suppressive effect of stress on natural killer cytotoxicity. *Endocrinology*, **126**, 2837–44.

70 Lukomska, B., Olszewski, W. L., Engeset, and Kolstad, P. (1983). The effect of surgery and chemotherapy on blood NK cell activity in patients with ovarian cancer. *Cancer*, **51**, 465–9.

71 Vinzenz, K., Matejka, M., Watzek, G., Poteder, H., Neuhold, N.,

and Micksche, M. (1987). Modulation of NK activity in regional lymph nodes by preoperative immunotherapy with OK-432 in patients with cancer of the oral cavity. *Cancer Detect. Prev. Suppl.*, **1**, 463–75.

72 Uchida, A., Kolb, R., and Micksche, M. (1982). Generation of suppressor cells for natural killer activity in cancer patients after surgery. *J. Natl. Cancer Inst.*, **68**, 735–41.

73 Miyazaki, S., Akiyoshi, T., Arinaga, S., Koba, F., Wada, T., and Tsuji, T. (1983). Depression of the generation of cell-mediated cytotoxicity by suppressor cells after surgery. *Clin. Exp. Immunol.*, **54**, 573–9.

74 Baigrie, R. J., Lewis, C. E., Lamont, P. M., Morris, P. J., and McGee, J. O'D. Effect of surgery on the release of interferon gamma by peripheral blood mononuclear cells: an investigation at the single cell level using the reverse haemolytic plaque assay. *Clin. Immunol. Immunopathol.* (Submitted).

75 Pollock, R. E., Lotzova, E., and Stanford, S. D. (1989). Surgical stress impairment of murine natural killer cell cytotoxicity involves pre- and post-binding events. *J. Immunol.*, **143**, 3396–403.

76 Urch, A., Muller, C., Aschauer, H., Resch, F., and Zielinski, C. C. (1988). Lytic effector cell function in schizophrenia and depression. *J. Neuroimmunol.*, **18**, 291–301.

77 Irwin, M., Daniels, M., Risch, S. C., Bloom, E., and Weuner, H. (1988). Plasma cortisol and natural killer cell activity during bereavement. *Biol. Psychiatry,* **24**, 173–8.

78 Levy, S. M., Herberman, R. B., Whiteside, T., Sanzo, K., Lee, J., and Kirkwood, J. (1990). Preceived social support and tumour estrogen/progesterone receptor status as predictors of natural killer cell activity in breast cancer patients. *Psychosom. Med.*, **52**, 73–85.

79 Khansari, D. N., Murgo, A. J., and Faith, R. E. (1990). Effects of stress on the immune system. *Immunol. Today,* **5**, 170–5.

80 Fabris, N. and Provinciali, M. (1989). Hormones. In *Natural immunity* (ed. D. S. Nelson), pp. 306–47. Academic Press, London.

81 Matera, L., Ciccarelli, E., Cesano, A., Veglia, F., Miola, C., and Camanni, F. (1989). Natural killer activity in hyperprolactinaemic patients. *Immunopharmacology,* **18**, 143–6.

82 Gerli, R., Rambotti, P., Nicoletti, I., Orlandi, S., Migliorati, G., and Riccardi, C. (1986). Reduced number of natural killer cells in patients with pathological hyperprolactinaemia. *Clin. Exp. Immunol.*, **64**, 399–406.

83 Nair, M. P. N. and Schwartz, S. A. (1984). Immunomodulatory effects of corticosteroids on natural killer and antibody dependent cytotoxic activities of human lymphocytes. *J. Immunol.*, **132**, 2876–82.

84 Onsurd, M. and Thorsby, E. (1981). Influence of *in vivo* hydrocortisone

on some human blood lymphocyte subpopulations. *Scand. J. Immunol.*, **13**, 573–9.

85 Hellstrand, K., Hermodsson, S., and Strannegard, O. (1985). Evidence for a β-Adrenoreceptor-mediated regulation of human natural killer cells. *J. Immunol.*, **134**, 4095–9.

86 Wiltrout, R. H. and Hornung, R. L. (1988). Natural products as antitumour agents: direct versus indirect mechanisms of activity of flavanoids. *J. Natl. Cancer Inst.*, **80**, 21–3.

87 Cummings, J. and Smyth, J. F. (1989). Flavone 8-acetic acid: our current understanding of its mechanism of action in solid tumours.

88 Ching, L. and Baguley, B. C. (1987). Induction of natural killer cell activity by the anti-tumour compound flavone acetic acid. *Eur. J. Cancer Clin. Oncol.*, **23**, 1047–50.

89 Wiltrout, R. H., Boyd, M. R., Back, T. T., Salup, R. R., and Hornung, R. L. (1988). Flavone-8-acetic acid augments systemic natural killer cell activity and synergizes with interleukin-2 for treatment of murine renal cancer. *J. Immunol.*, **140**, 3261–5.

90 Damia, G., Tagliabue, G., Allavena, P., and D'Incalci, M. (1990). Flavone acetic acid antitumour activity against a mouse pancreatic adenocarcinoma is mediated by natural killer cells. *Cancer Immunol. Immunother.*, **32**, 241–4.

91 Ghosh, A. K., Mellor, M., Prendville, J., and Thatcher, N. (1990). Recombinant interleukin-2 (rIL-2) with flavone acetic acid (FAA) in advanced malignant melanoma: immunolgical studies. *Br. J. Cancer*, **61**, 471–4.

92 Ching, L. M. and Baguley, B. C. (1989). Reduction of cytotoxic cell activity in colon 38 tumours following treatment with flavone acetic acid. *Eur. J. Cancer Clin. Oncol.*, **25**, 1061–5.

93 Gasic, G., Gasic, T., Galanti, N., Johnson, T., and Murphy, S. (1973). Platelet-tumour cell interactions in mice: the role of platelets in the spread of malignant disease. *Int. J. Cancer*, **11**, 704–11.

94 Rickles, R. and Edwards, R. (1983). Activation of blood coagulation in cancer: Trouseau's syndrome revisited. *Blood*, **62**, 14–19.

95 Timonen, T., Ortaldo, J. R., and Herberman, R. B. (1982). Nalysis by a single cell cytotoxicity assay of natural killer (NK) cell frequencies among human large granular lymphocytes, and of the effects of interferon on that activity. *J. Immunol.*, **128**, 2514–21.

96 Taniguchi, K., Peterson, M., Hoglund, P., Kiessling, R., Klein, G., and Karre, K. (1987). Interferon gamma induces lung colonization by intravenously inoculated B16 melanoma cells in parallel with enhanced expression of class I major histocompatibility complex antigens. *Proc. Natl. Acad. Sci. U.S.A.*, **84**, 3405–9.

97 Ferrat. N. J., *et al.* (1990). Recombinant gamma interferon provokes resistance of human breast cancer cells to spontaneous and

IL-2 activated non-MHC restricted cytotoxicity. *Br. J. Cancer,* **61,** 558–62.

98 Faith, R. E., Liang, H. J., Plotnikoff, N. P., Murgo, A. J., and Nimeh, N. F. (1987). Neuroimmunomodulation with enkephalins: *in vitro* enhancement of natural killer cell activity in peripheral blood lymphocytes from cancer patients. *Nat. Immunol. Cell Growth Regul.,* **6,** 88–98.

99 Kay, N. E., Morley, J. E., and Allen, J. I. (1990). Interaction between endogenous opioids and IL-2 on PHA-stimulated human lymphocytes. *Immunology,* **70,** 485–9199.

100 Lefor, A. T., Mule, J. J., and Rosenberg, S. A. (1989). Lymphokine-activated killer cells: biology and therapeutic efficacy. In *Functions of the natural immune system* (ed. C. W. Reynolds and R. H. Wiltrout). Plenum, New York.

101 Tilden, A. B., Itoh, K., and Balch, C. M. (1987). Human lymphokine-activated killer cells: identification of two types of effector cells. *J. Immunol.,* **138,** 1068–73.

102 Ortaldo, J. R. (1989). Cytokine production by CD3⁻ large granular lymphocytes. In *Functions of the natural immune system* (ed. C. W. Reynolds and R. H. Wiltrout), pp. 299–311. Plenum, New York.

103 Ruggiero, V., Latham, K., and Baglioni (1987). Cytostatic and cytotoxic activity of tumour necrosis factor on human cancer cells. *J. Immunol.,* **137,** 2711–16.

104 Ortaldo, J. R., Ransom, J. R., Sayers, T. J., and Herberman, R. B. (1986). Analysis of cytostatic/cytotoxic lymphokines: relationship of natural killer cytotoxic factor to recombinant lymphotoxin, recombinant tumour necrosis factor on human cancer cells. *J. Immunol.,* **137,** 2857–62.

105 Chong, A. S-F., Scuderi, P., Grimes, W. J., and Hersh, E. M. (1989). Tumour targets stimulate activated killer cells to produce interferon-gamma and tumour necrosis factor. *J. Immunol.,* **142,** 2133–9.

106 Ramani, P. and Balkwill, F. R. (1987). Enhanced metastasis of a mouse carcinoma after *in vitro* treatment with murine interferon gamma. *Int. J. Cancer,* **40,** 830–4.

107 McMillan, T. J., Rao, J., Everett, C. A., and Hart, I. R. (1987). Interferon-induced alterations in metastatic capacity, class 1 antigen expression and natural killer cell sensitivity of melanoma cells. *Int. J. Cancer,* **40,** 659–63.

108 Leibovich, S. J., Polverini, P. J., Shepherd, H. M., Weiseman, D. M., Shively, V., and Nuseir, N. (1987). Macrophage-induced angiogenesis is mediated by tumour necrosis factor alpha. *Nature,* **329,** 630–2.

109 Malik, S. T. A., Griffen, D. B., Fiers, W., and Balkwill, F. R. (1989). Paradoxical effects of tumour necrosis factor in experimental ovarian cancer. *Int. J. Cancer,* **44,** 918–25.

110 Malik, S. T. A., Naylor, M. S., East, N., Oliff, A., and Blakwill, F. R. (1990). Cells secreting tumour necrosis factor show enhanced metastasis in nude mice. *Eur. J. Cancer*, **26**, 1031–4.

111 Migliori, R. J., *et al.* (1987). Lymphokine-activated killer (LAK) cells can be focussed at sites of tumour growth by products of macrophage activation. *Surgery*, **102**, 155–62.

112 Pillai, M. R., Balaram, P., Abraham, T., Padmanabhan, T. K., and Nair, M. K. (1988). Natural cytotoxicity and serum blocking in malignant cervical neoplasia. *Am. J. Reprod. Immunol. Microbiol.*, **16**, 159–62.

113 Guillou, P. J., Sedman, P. C., and Ramsden, C. W. (1989). Inhibition of lymphokine-activated killer cell generation by cultured tumour cell lines in vitro. *Cancer Immunol. Immunother.*, **28**, 43–53.

114 Kawakami, Y., Custer, M. C., Rosenberg, S. A., and Lotze, M. T. (1989). IL-4 regulates IL-2 induction of lymphokine-activated killer activity from human lymphocytes. *J. Immunol.*, **142**, 3452–61.

115 Rosenberg, S. A. (1990). Adoptive immunotherapy for cancer. *Sci. Amer.*, May, 34–41.

116 Schwartz, R. E., Iwatsuki, S., Herberman, R. B., and Whiteside, T. L. (1989). Unimpaired ability to generate adherent lymphokine-activated killer (A-LAK) cells in patients with primary or metastatic liver tumours. *Cancer Immunol. Immunother.*, **30**, 312–16.

117 Bosnes, V. and Hirschberg, H. (1989). Immunomagnetic separation of infiltrating T lymphocytes from brain tumours. *J. Neurosurg.*, **71**, 218–23.

118 Topalain, S. L., Muul, L. M., Solomon, D., and Rosenberg, S. A. (1987). Expansion of human tumour infiltrating lymphocytes for use in immunotherapy trials. *J. Immunol. Methods*, **102**, 127–41.

119 Russell, S. J. (1990). Lymphokine gene therapy for cancer. *Immunol. Today*, **11**, 196–9.

120 Domzig, W. and Stadler, B. M. (1982). The relationship between human natural killer cells and interleukin 2. In *NK cells and other natural killer cells* (ed. R. B. Herberman), pp. 409–20. Academic Press, New York.

121 Procopio, A. D., Allavena, P., and Ortaldo, J. R. (1985). Noncytotoxic functions of natural killer (NK) cells. Large granular lymphocytes (LGL) produce a B cell growth factor (BCGF). *J. Immunol.*, **135**, 3264–71.

122 Kasahara, T., Djeu, J. Y., Dougherty, S. F., and Oppenheim, J. J. (1983). Capacity of human large granular lymphocytes (LGL) to produce multiple lymphokines: interleukin 2, interferon and colony stimulating factor. *J. Immunol.*, **131**, 2379–85.

123 Blanca, I., Herberman, R. B., and Ortaldo, J. R. (1985). Human natural killer cytotoxic factor. Studies on its production, specificity

and mechanism of interaction with target cells. *Nat. Immun. Cell Growth Regul.*, **4**, 48–59.

124 Herberman, R. B., Reynolds, C. W., and Ortaldo, J. R. (1985). Mechanism of cytotoxicity by natural killer cells. *Ann. Rev. Immunol.*, **4**, 651–80.

125 Paciotti, G. F. and Tamarkin, L. (1988). Interleukin-2 differnatially affects the proliferation of a hormone dependent and a hormone-independent human breast cancer cell line *in vitro* and *in vivo*. *Anticancer Res.*, **8**, 1233–40.

126 Sayers, T. J., Ransom, J. R., Denn, A. C., Herberman, R. B., and Ortaldo, J. R. (1986). Analysis of a cytostatic lymphokine produced by incubation of lymphocytes with tumour cells: relationship to leukoregulin and distinction from recombinant lymphotoxin, recombinant tumour necrosis factor, and natural killer cytotoxic factor. *J. Immunol.*, **137**, 385–90.

127 Platsoucas, C. D., *et al.* (1989). Regulation of natural killer cytotoxicity by recombinant alpha interferons. Augmentation by IFN-α7, and interferon similar to IFN-α. *J. Anticancer Res.*, **9**, 849–58.

128 Ortaldo, J. R., Phillips, W., Wasserman, K., and Herberman, R. B. (1980). Effects of metabolic inhibitors on spontaneous and interferon-boosted human natural killer cell activity. *J. Immunol.*, **134**, 794–802.

129 Sayers, T. J. A., Mason, A. T., and Ortaldo, J. R. (1986). Regulation of human natural killer cell activity by interferon-gamma: lack of a role in interleukin-2 mediated augmentation. *J. Immunol.*, **136**, 2176–82.

130 Henney, C. S., Kuribayashi, K., Kern, D. E., and Gillis, S. (1981). Interleukin-2 augments natural killer cell activity. *Nature*, **291**, 335–6.

131 Lattime, E. C., Percoraro, G. A., and Stutman, O. (1983). The activity of natural cytotoxic cells is augmented by interleukin-2 and interleukin-3. *J. Exp. Med.*, **157**, 1070–80.

132 Nagler, A., Lanier, L. L., and Phillips, J. H. (1988). The effects of IL-4 on human natural killer cells. *J. Immunol.*, **141**, 2349–51.

133 Luger, T. A., *et al.* (1989). IFN β2/IL-6 augments the activity of human natural killer cells. *J. Immunol.*, **143**, 1206–9.

134 Voth, R., *et al.* (1988). *In vivo* and *in vitro* induction of natural killer cells by cloned human tumour necrosis factor. *Cancer Immunol. Immunother.*, **27**, 128–32.

135 Leung, K. H. and Koren, H. S. (1986). Regulation of human natural killing. II. Protective effect of interferon on NK cells from suppression by PGE_2. *J. Immunol.*, **129**, 1742–7.

136 Sone, S., Utsugi, T., Nii, A., and Ogura, T. (1988). Differential effects of recombinant interferons alpha, beta and gamma on induction of human lymphokine (IL-2)-activated killer activity. *J. Natl. Cancer Inst.*, **80**, 425–31.

137 Dunlap, N. E., Lane, V. G., Cloud, G. A., and Tilden, A. B. (1990). *In vitro* natural killer and lymphokine-activated killer in activity in patients with broncogenic carcinoma. *Cancer,* **66,** 1499–1504.

138 Gallagher, G., Al-Azzawai, F., Davis, J., and Stimson, W. H. (1989). Cytokines regulate the ability of human LAK-cells to kill human tumour cells *in vitro. Br. J. Cancer,* **59,** 919–21.

139 Lewis, C. E., Ramshaw, A. L., Lorenzen, J., and McGee, J. O'D. (1991). Basic fibroblast growth factor and interleukins 4 and 6 stimulate the release of IFN-gamma by individual NK cells. *Cell. Immunol.,* **132,** 158–67.

7 Natural killer cells/large granular lymphocytes in pregnancy

P. M. STARKEY

1 Introduction

One of the central paradoxes of human pregnancy is that the genetically distinct fetus is able to survive and develop within the mother, without provoking an immune rejection response. The placenta and fetal membranes are the fetal tissues in most intimate contact with the maternal decidua which lines the pregnant uterus, and it is presumed that the interaction between the decidua and these fetal tissues is crucial to the success of pregnancy. The anatomy of the human pregnant uterus is depicted schematically in Fig. 7.1.

The surface of the placenta, which is made up of innumerable branched villi, is covered by the multinucleate syncytium of syncytiotrophoblast, which is in contact with the maternal blood. Across the syncytiotrophoblast,

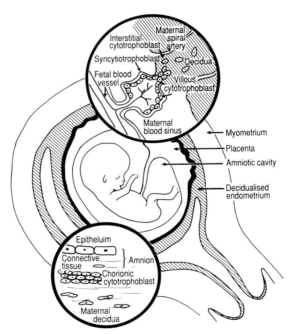

Fig. 7.1 A schematic diagram of the tissues of the pregnant human uterus. The whole uterus with the fetus *in situ* is depicted, with two areas enlarged and shown in more detail. At the point of attachment of the placenta, villous cytotrophoblast are shown underneath the syncytiotrophoblast which covers the surface of the placenta and is in contact with the maternal blood sinuses. Extra-villous or interstitial cytotrophoblast invade the decidua and anchor the placental villi. The second enlargement shows the cell layers of the fetal membranes, or amniochorion, and attached decidua.

blood gases and nutrients diffuse to the fetus and fetal waste products to the maternal circulation. The placenta is anchored in the decidua, however, by the cytotrophoblast which invade the decidua from the tips of the chorionic villi. Cytotrophoblast are also the fetal cell type in contact with the maternal decidua in the fetal membranes or amniochorion, and are responsible for the initial invasion of the decidua resulting in implantation of the early blastocyst, and the invasion of the uterine spiral arteries that occurs in the first four months of pregnancy. This trophoblast invasion of the uterine arteries depicted in Fig. 7.2, is associated with degeneration of the vascular endothelium and musculoelastic tissue. This converts the artery into a flaccid tube which can subsequently expand as pregnancy progresses. If this process does not occur, the arteries cannot accommodate the hugely increased maternal blood flow that is necessary for the placenta and fetus to continue to grow.

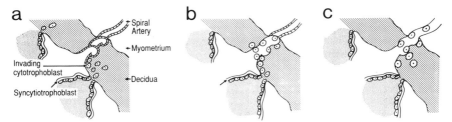

Fig. 7.2 The process of invasion of the maternal spiral arteries by fetal cytotrophoblast. (a) Initial stages, with the muscular walls of the spiral arteries intact and cytotrophoblast beginning to invade the decidual stroma from the anchoring villi. (b) Invasion by cytotrophoblast of the part of the artery that is within the decidua and erosion of the vessel wall so that it can expand to carry more blood. The section of the artery in the myometrium is still narrow. (c) A second wave of trophoblast invasion of the myometrial section of the artery, causing further widening. It is this last stage which fails in pre-eclampsia, preventing the expansion of the maternal blood flow later in pregnancy (129).

Thus, at several stages the development of the placenta, and the success of the pregnancy, depends on the ability of cytotrophoblast to infiltrate the maternal decidua. Failure at the blastocyst stage blocks implantation, or may subsequently result in spontaneous abortion. Failure of trophoblast invasion of the uterine arteries is associated with both pre-eclampsia, a condition responsible for increased fetal and maternal morbidity and mortality, and intra-uterine growth retardation which is associated with an increased rate of fetal handicap (1). Conversely, excessive trophoblast invasion can lead to placenta accreta with an increased risk of maternal haemorrhage.

The only fetal cells in direct contact with maternal blood or tissue are the syncytiotrophoblast and cytotrophoblast. Syncytiotrophoblast expresses neither Class I nor Class II MHC antigens (2) and is, therefore, unlikely to provoke a conventional T-cell-mediated rejection response. It seems that the bulk of the placenta presents a barrier between the fetus and the maternal immune system, with an 'immunogenically inert' surface. The cytotrophoblast on the other hand, which are in intimate contact with the maternal decidua, do express MHC Class I, although in the form of the apparently monomorphic HLA-G (3). There is some evidence that cyto-trophoblast may also express an unusual form of MHC Class II (4).

In human pregnancy, the placenta may survive and develop within the maternal tissues, partly because it does not express polymorphic MHC antigens. In other species, however, including mouse and horse, some trophoblast populations do express polymorphic MHC Class I antigens (5, 6), and even in human pregnancy, minor histocompatibility antigens or other polymorphic membrane proteins on trophoblast may be recognized as non-self by the maternal immune cells, precipitating a local immune or inflammatory response. Attention has, therefore, focused on the cells present in the maternal decidua, and the way in which they might regulate the maternal response to trophoblast.

The decidua develops from the endometrium, which in the non-pregnant uterus undergoes cyclical changes, some of which prefigure the changes which occur after conception. The non-pregnant endometrium and the decidua contain the same cell populations, but the relative proportions of these cell types, and their functional activity, vary during the menstrual cycle and in pregnancy. The decidua consists of fibroblast-derived stromal cells embedded in an extracellular matrix that they secrete. The tissue is vascularized by the uterine vessels and the luminal surface is covered by columnar epithelial cells that also line the numerous glands which extend through the whole decidua. During the second trimester these glands regress and have virtually disappeared at term (7). However, the decidua also contains numerous cells of bone marrow origin and, in early pregnancy, the majority of these express antigens characteristic of NK cells or large granular lymphocytes (LGLs). It is the characteristics and possible function(s) of these unusual decidual LGLs which will be the subject of this chapter. These will be referred to as LGLs rather than NK cells, as their NK cell activity is relatively weak, and it is probable that their function *in vivo* is not simply limited to their cytotoxic activity.

2   LGLs at the materno-fetal interface

2.1   Cell populations in decidua in early pregnancy

The demonstration of cells of bone marrow origin in human pregnancy decidua has relied on studies of cell surface antigen expression. Immunohistology of first trimester decidua showed numerous cells expressing the leucocyte common antigen (CD45) (8). A majority of these cells were CD56$^+$ (9) with the morphology of large lymphocytes and were found scattered throughout the decidua or in clumps of lymphoid cells, often in association with T-cells (see Fig. 7.3). There is circumstantial evidence (8, 10, 11) that the CD56$^+$ cells are identical to the so-called 'endometrial granulocytes'. The latter were defined by classical histochemistry on the basis of their strongly acidophilic granules which stain with phloxine-tartrazine (12).

Macrophages, which are positive for MHC Class II (13), are found throughout the decidua including around the spiral arteries. They have a stellate morphology with long processes interdigitating between other decidual cells, including the aggregates of T-cells and LGLs. B-cells are very rare in such tissues.

More quantitative assessment of decidual cell populations in early pregnancy has been achieved by flow cytometry. Decidual cell dispersions, prepared by enzymic digestion of first trimester tissue, were labelled with various combinations of fluorescein- and phycoerythrin-conjugated antibodies to surface antigens (14). Using this technique about 75 per cent of decidual cells were found to be CD45$^+$. Cells expressing antigens characteristic of peripheral blood LGLs made up 45 per cent of decidual cells, 19 per cent were macrophages, and 8 per cent T-cells. The decidual macrophages were positive for MHC Class II and CD11c, and the T-cells expressed CD3, CD2, and CD5 but were largely negative for MHC Class II.

Decidual LGLs, however, were found to have an unusual antigen expression (14) and could be divided into three distinct subsets (Table 7.1), although all were positive for CD38 and negative for MHC Class II. Most decidual LGLs (subsets 1a and 1b) were strongly positive for CD56 but negative for CD16, the Fc receptor (FcтRIII) found on peripheral blood LGLs. These cells were also negative for CD57 and for various T-cell markers including CD5 and CD3. About half of the CD56$^+$/CD16$^-$ LGLs also expressed CD2, the E rosette receptor, and CD7 (subset 1b), whereas the remainder were CD2$^-$ and CD7$^-$ (subset 1a) (14). Immunohistochemical studies using double labelling with peroxidase-conjugated anti-CD56 and alkaline phosphatase-conjugated anti-CD2 confirmed that the CD56$^+$/CD2$^-$ subset was not an artefact resulting from the action of the enzymes used to prepare the cell dispersion (P. M. Starkey, unpublished observations). The CD56$^+$ decidual LGLs were negative for CD25, the p55 protein of the IL-2 receptor but positive for the p75 IL-2 receptor and for the IL-4 receptor (134).

A minor subset of decidual LGLs (subset 2) were negative or only dimly positive for CD56, but positive for CD16 and negative for CD2 and CD3. These cells resemble the major type of LGLs in peripheral blood, the CD16^{bright} CD56^{dim} subset (15). In contrast the major decidual LGL type,

(a)

(b)

(c)

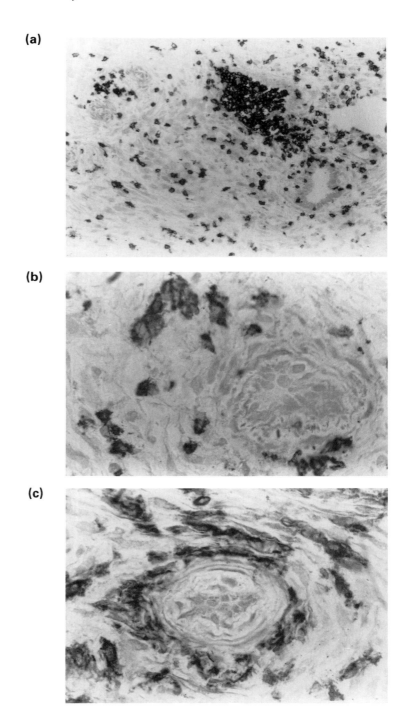

Fig. 7.3 Immunohistology of first trimester decidua demonstrating CD56$^+$ LGLs and decidual macrophages. Frozen tissue sections of decidua taken from first trimester termination material, were fixed in acetone, and then stained with a mouse monoclonal to CD56 (NKH1 from Coulter Immunology, Luton, Beds, UK), or with 3C10 a macrophage-specific marker (27), followed by rabbit anti-(mouse Ig) Ig and a preformed complex of mouse anticalf intestinal alkaline phosphatase and alkaline phosphatase. Bound antibody was visualized as red staining (131) and sections were counterstained with haematoxylin. (a) CD56$^+$ LGLs scattered throughout the decidual stroma, and concentrated in a lymphoid aggregate. A decidual gland is visible near the bottom right, and blood vessels near the top left ($\times 400$). (b) High power view of a blood vessel in the decidua with CD56$^+$ LGLs in the adjacent stroma ($\times 1600$). (c) 3C10$^+$ macrophages around a blood vessel in decidual stroma. Unlike LGLs, decidual macrophages are clustered in a perivascular location as well as being scattered through the stroma ($\times 1600$).

---

the CD16$^-$/CD56$^+$ cells, correspond most closely to the minor populations, comprising less than 10 per cent of peripheral blood LGLs, the CD16^{dim}/CD56^{bright} and CD16$^-$/CD56^{bright} subsets (15). The CD16$^-$/CD56$^+$ and CD16$^+$/CD56$^-$ subsets of decidual LGLs have been isolated separately by flow cytometry to purities of greater than 94 per cent (14). The flow cytometric profiles of the purified LGLs are shown in Fig. 7.4. The CD16$^+$/CD56$^-$ cells are larger and more granular than the CD16$^-$/CD56$^+$ cells, and the latter appeared to contain two populations on the basis of granularity (14) confirming the results of Giemsa staining of cell dispersions which identified CD56$^+$ LGLs as a mixture of granular and agranular cells (11).

Clones derived from decidual CD3$^-$ cells have recently been isolated (16). Decidual cells, depleted of cells positive for the α/β T-cell receptor, were stimulated with IL-2 and phytohaemagglutinin (PHA) and the proliferating cells cloned by limiting dilution and grown with IL-2. Most of the CD3$^-$ clones were CD56$^+$ and about a third of these were also CD16$^+$. Southern blot analysis demonstrated that in all CD3$^-$ clones tested, the T-cell receptor γ and δ genes were in the germ line configuration. The proliferative frequency of the decidual cells in these cloning studies was very low, that is, 1 in 200, compared with that of peripheral blood, and the clones obtained were not fully representative of the LGL types present in decidua. Nevertheless, despite the over-representation of CD56$^+$/CD16$^+$ cells, the authors believe their clones were derived from decidual LGLs and not from any contaminating blood LGLs.

Bone marrow-derived cells have also been described in pregnancy decidua of non-human species. In the mouse, the major decidual population in the first half of pregnancy, 45–80 per cent of the total, has a 'null cell' phenotype, displaying neither B- nor T-cell markers (17, 18) and probably corresponds to the LGLs of human decidua. B-cells, T-cells, and macrophages are also present, with the proportion of T-cells and macrophages

Table 7.1 Surface antigen expression of human
decidual LGL subsets.

Antigen	Decidual LGL subset		
	1a	1b	2
CD56	+	+	−
CD16	−	−	+
CD2	−	+	−
CD7	−	+	−
CD57	−	−	−
CD38	+	+	+
CD3	−	−	−
CD5	−	−	−
MHC Class II	−	−	−
IL-2R			
CD25	−	−	NT
p75	+	+	NT
IL-4R	+	+	NT

The data is compiled from flow cytometric analysis of human
decidual cell dispersions (14, 15). NT, not tested.

increasing with gestational age (18, 19). Rodents also have a specialized
area of the decidua called the metrial gland, which develops during the
second half of pregnancy at each implantation site (20). The metrial gland
cells are granulated lymphocytes, which have been shown in chimeric mice
to be derived from the bone marrow (21).

Leucocyte invasion of the decidua has also been described in the horse,
but this is associated with the formation of the endometrial cups, a phenomenon unique to the horse and related species. In early pregnancy in the
horse, the spherical conceptus becomes ringed by a girdle of highly invasive cytotrophoblast. These chorionic girdle cells destroy the endometrial epithelium, pass down the glands, and enter the endometrial
stroma forming the so-called endometrial cups. The endometrial cup cells
are the source of equine chorionic gonadotrophin until the end of the first
trimester. Maternal leucocytes accumulate in the endometrium around the
endometrial cups, apparently walling them off. Later on, as the endometrial cup cells degenerate, the maternal leucocytes invade the cups. In
contrast to human and murine decidual cells, the equine endometrial
leucocytes have been shown to comprise about 50 per cent T-cells with
some macrophages and the rest null cells (6).

Bone marrow-derived cells have also been detected in the sheep uterus,
although the only study has been of non-pregnant endometrium (22). Most
were granular lymphocytes lacking B- and T-cell markers. There was no
obvious variation in populations with the stage of the oestrus cycle.

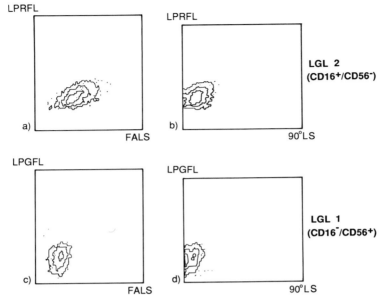

Fig. 7.4 Flow cytometric profiles of purified decidual LGL subsets. Cell dispersions from first trimester decidua were labelled with phycoerythrin-conjugated anti-CD16, or anti-CD56 followed by fluorescein-conjugated second antibody. The antibody-positive cells were selected by flow cytometry and the pure cell populations re-analysed. Cell populations were either CD16$^+$/CD56$^-$ (a,b) or CD16$^-$/CD56$^+$ (c,d) and were analysed on a Coulter EPICS 541 flow cytometer for fluorescence intensity (log peak red or green fluorescence—LPRFL or LPGFL) and forward angle light scatter (FALS) a measure of size (a,c) or fluorescence intensity and 90° light scatter (90°LS) a measure of cell granularity (b,d). (reproduced with permission from (14).)

2.2 Variation in LGL populations in non-pregnant endometrium and during pregnancy

In early human pregnancy decidua, LGLs are the most abundant cell type, but this is not the case throughout pregnancy, nor in non-pregnant endometrium for most of the time. Morphometric assessment using antibody-labelled tissue sections, of the proportions of each cell type present in non-pregnant endometrium (Fig. 7.5) has demonstrated that the number of macrophages and T-cells is more or less constant throughout the menstrual cycle, apart from an increased density of macrophages during menstruation. There are negligible numbers of B-cells (11, 23). In the latter half of the menstrual cycle, however, the number of CD56$^+$ cells increases sharply and they become the dominant cell type. At around the

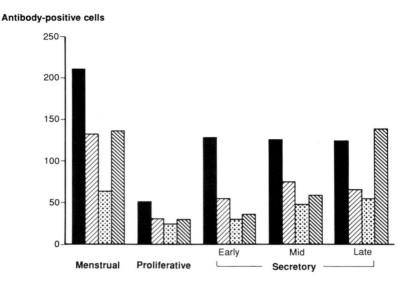

Fig. 7.5 Variation throughout the menstrual cycle of bone marrow-derived cell populations in human endometrium. Frozen tissue sections of non-pregnant endometrium taken throughout the menstrual cycle from healthy pre-menopausal women were labelled with monoclonal antibodies to CD45 (■), 3C10, a macrophage marker (27), (▨), CD5 for T-cells (▦), and CD56 for LGLs (▧) followed by a peroxidase-conjugated second antibody. The number of positive cells in 1562.5 μm^2 of tissue was determined morphometrically. Tissues were grouped according to stage in the cycle, and median values taken. (Data taken from (24).)

expected time of implantation CD56$^+$ LGLs constitute 70–80 per cent of the bone marrow-derived endometrial cells (11, 23). Numbers increase still further if conception occurs, to give the levels of CD56$^+$ cells found in early pregnancy decidua (11). If conception does not occur, the nuclei of the CD56$^+$ cells become pyknotic suggesting cell death, and the proportion of LGL drops (11).

As pregnancy progresses beyond the first trimester the numbers of decidual LGLs decline dramatically. Immunohistological studies of the implantation site in the second and third trimester of human pregnancy suggest that most of the CD45$^+$ decidual cells were macrophages, with only a few CD3$^+$/CD2$^+$ T-cells (24, 25). Many of the T-cells were also CD8$^+$ (25). Flow cytometric analysis of cell dispersions of term decidua (26) showed, on average, 47 per cent of decidual cells to be CD45$^+$, comprising only 4 per cent CD56$^+$ LGLs, with 18–19 per cent positive for MHC Class II and for CD14, a macrophage-specific marker (27) and 8 per cent to be CD3$^+$ T-cells. The remaining cell population, which was very variable between individuals, comprised polymorphonuclear leucocytes (Fig. 7.6).

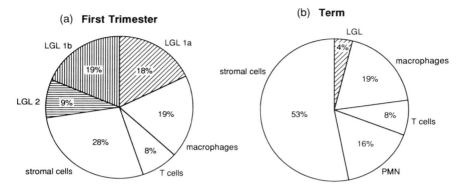

Fig. 7.6 Cell populations in human decidua in the first trimester of pregnancy and at term. Cell dispersions of decidual tissue were labelled with fluorescent antibodies and analysed by flow cytometry. (a) First trimester, the median values for 15 samples, gestational age 7–13 weeks. LGL1a, CD16$^-$/CD56$^+$/CD2$^-$; LGL1b, CD16$^-$/CD56$^+$/CD2$^+$; LGL2, CD16$^+$/CD56$^-$; macrophages, HLA Class II$^+$; T-cells, CD3$^+$; stromal cells, CD45$^-$. (Data taken from (14).) (b) Term, the mean values for 19 samples. LGL, CD56$^+$; macrophages, HLA Class II$^+$; T-cells, CD3$^+$; PMN, CD16$^+$ detected with 3G8 an antibody to the PMN Fcγ receptor (131) and verified morphologically by Giemsa staining of cytospins. (Data taken from (26).)

2.3   Factors regulating LGL numbers in the decidua

Considerable evidence suggests that the changes which occur in human decidua during early pregnancy are largely under hormonal control, rather than being a direct response to the presence of trophoblast. Circumstantial evidence for this is provided by the fact that the changes during the latter half of the menstrual cycle prefigure the early pregnancy changes, and only persist if progesterone levels remain elevated (28). More direct evidence comes from studies of intra-uterine decidua in ectopic pregnancies. Glandular changes (29), infiltration by bone marrow-derived cell populations (P. M. Starkey, L. M. Clover, and B. McCallum, unpublished observations) and inhibition of prostaglandin synthesis (30) were all similar in intra-uterine decidua from either normal or ectopic pregnancies. The only difference appeared to be that decidual blood vessels were less enlarged in ectopic pregnancy (31).

In contrast, at the site of ectopic implantation in the fallopian tube, although classical histology detected focal 'decidualization' of the tubal mucosa in some cases (29, 32), the bone marrow-derived cells infiltrating the tubal tissue around the conceptus were those normally associated with inflammation (29). No evidence was found of CD56$^+$ LGLs (P. M. Starkey, *et al.*, unpublished observations).

The evidence from ectopic pregnancies suggests that the increase in

CD56$^+$ LGLs is specific to the uterus. Thus, the hormonal effect must be mediated via another decidual cell type, such as the T-cells, stromal or glandular cells. In endometrium, progesterone and oestrogen receptors have been detected on the glandular and stromal cells, but oestrogen receptors have also been demonstrated immunohistologically on a sub-population of CD3$^+$ T-cells found in the lymphoid aggregates (23).

In rodents, the decidual changes occur, provided that the uterus has been hormonally prepared, either as a result of implantation, or in pseudopregnancy as a result of non-specific stimuli (28). The metrial gland only develops after changes in the decidual itself, suggesting that the signal for its differentiation originates in the decidua, and may be a single or number of cytokines produced by decidual cells (28).

There is evidence from immunohistology of human non-pregnant endometrium and first trimester pregnancy decidua that some at least of the increase in CD56$^+$ LGLs is due to proliferation *in situ*. In non-pregnant endometrium, the antibody Ki67 which binds to proliferating cells (34) labels 20–30 per cent of the CD45$^+$ cells in the mid- to late-secretory phase, including some CD56$^+$ cells. Similar results were obtained using incorporation of bromodeoxyuridine into S-phase cells (35, 36). In early decidua, however, although the numbers of CD56$^+$ cells were much higher than in non-pregnant endometrium (11), fewer cells of lymphocyte morphology were labelled with Ki67 (35, 37).

In the mouse, the total numbers of decidual cells continue to increase well into the second half of pregnancy (18), and several lines of evidence suggest that this increase is due to the migration of cells from the bone marrow to the decidua via the uterine lymph nodes. Injection of fluorophor-labelled bone marrow cells into the tail vein of virgin or pregnant mice results in preferential localization of labelled cells in the decidua and uterine lymph nodes of pregnant animals (38). Null cell numbers increase first in the bone marrow, then in the blood and finally in the uterine lymph nodes. These increases are more marked in allogeneic than in syngeneic pregnancies suggesting that the interaction between the conceptus and the decidua has some effect on the strength of the decidual response (39).

It is clear that in both mouse and human pregnancy, although some proliferation may occur *in situ*, CD56$^+$ LGLs home in on the decidua. This cannot be a direct response to increased systemic progesterone levels nor, since it also occurs in ectopic pregnancies, to a trophoblast product, but is most likely due to the production of an attractant molecule by decidual T-cells, stromal or glandular cells. In this context, it is noteworthy that peripheral blood LGLs that most closely resemble decidual LGL, (that is, the CD16^{dim}/CD56^{bright} and CD16$^-$/CD56^{bright} subsets), as well as expressing much higher levels of CD56 than the majority of blood LGL (40), also express higher levels of CD11c and CD44 (15). All three of these antigens are involved in cell adhesion. CD56 has recently been shown to be an isoform of the neural cell adhesion molecule (NCAM) (41). NCAM molecules associate as homodimers, and the CD56 of LGLs plays a role in the cytotoxic activity of LGL by binding to NCAM on the surface of some target cells (42). NCAM is not expressed in decidua apart from on decidual LGLs, but it is possible that CD56 recognizes other ligands. CD44 is the surface antigen involved in lymphocyte binding to high endothelial venules, and CD11c is the p150,95 protein, one of a family of leucocyte adhesion glycoproteins. Both CD44 and CD11c have been implicated in the migration across the vascular endothelium of lymphocytes and monocytes respectively (43, 44). It seems possible that any or all of these adhesion molecules might be involved in the migration of the precursors of decidual LGLs into the tissue.

3   Circulating LGLs in pregnancy

The most profound changes in LGL populations in pregnancy occur at the maternal/fetal interface in the uterus, but pregnancy also affects LGLs in the blood. Most studies have measured NK activity and compared blood samples from pregnant and non-pregnant women and from men. Blood NK activity was found to be lower in women than in men, and to fall significantly in non-pregnant women in the periovulatory period. There was no such drop in women on oral contraceptives (45). In pregnancy, NK activity of blood lymphocytes was decreased compared with non-pregnant controls, the reduction reaching statistical significance from 16 weeks of pregnancy onwards. NK activity returned to control levels between nine and 40 weeks post-partum (46). Others reported that antibody-dependent cell-mediated cytotoxic (ADCC) activity decreases during pregnancy and returns to slightly above control levels post-partum (47).

The mechanism for the reduction in peripheral NK activity in pregnancy is not clear. The CD16[+]/CD57[+] subset of blood lymphocytes decreases in pregnancy, whereas the CD16[+]/CD57[−] population is unaltered (48). Interestingly, the opposite change was reported in renal transplant patients treated with immunosuppressive drugs (49). Using a single cell cytoxicity assay, it was shown that the reduced NK activity in pregnancy correlated with a reduction in the ability of lymphocytes to kill NK cell-sensitive target cells, though the percentage of lymphocytes binding to the targets was the same in pregnant and non-pregnant samples (50). Lymphocytes from non-pregnant subjects incubated *in vitro* in serum from pregnant women also showed reduced lytic activity compared to cells incubated in pooled non-pregnancy serum (51). The effect may be mediated by sex steroid hormones, since oestrogen is known to reduce NK activity in mice (52), and luteinizing hormone and human chorionic gonadotrophin (hCG) at physiological concentrations inhibit NK activity *in vitro* (45).

Alterations in NK activity have been reported in abnormal pregnancies. Lymphocytes from women with threatened pre-term labour were shown to have increased cytotoxicity against NK cell targets or fetal fibroblasts compared with controls. The effect was due to increased lytic activity not increased binding to target cells (53). In pre-eclampsia, contradictory results have been reported. Toder and co-workers (54) found NK cytotoxicity to be increased in pre-eclamptics compared with controls, and showed that IFN did not activate NK activity *in vitro* of cells from pre-eclamptics unlike those from control pregnant women. In contrast, Alanen and Lassila (55) found cytotoxicity to be significantly decreased in pre-eclampsia, and using IFNα were able to increase NK activity in all samples.

4   Function of decidual LGLs

Investigation of the role of the maternal decidua in regulating placental development focused initially on the question of how the fetus, and more particularly the placenta, could survive in intimate contact with maternal tissues, given that it was genetically foreign.

There is much evidence that the pregnant uterus is an immunologically privileged site, and this has led to the suggestion that a prime role for the decidua is to down-regulate any local response of the maternal immune system to placental trophoblast. An immunosuppressive function of the decidua would have an essentially permissive role in regulating placental development; with normal placentation being compromised if the maternal immune response was not suppressed.

Alternatively, the decidua might play a more positive role by directly affecting trophoblast growth and development. Thus, decidua might inhibit the infiltration of trophoblast into the decidua and uterine arteries, limiting the extent of placentation, or conversely by 'immunotrophism' might promote placental growth and development by secreting factors which stimulate trophoblast.

The ways in which decidual LGLs might regulate placental growth are depicted schematically in Fig. 7.7. Some of these effects are contradictory, and the eventual outcome for the placenta may be seen as a balance between opposing regulatory forces.

Modulation by decidual cells of the maternal immune response, or trophoblast function could be mediated either by direct cell contact or indirectly through the secretion of soluble factors including cytokines. Relatively little is known at present about the cytokines produced by decidual cells, but much current research is concerned with this topic.

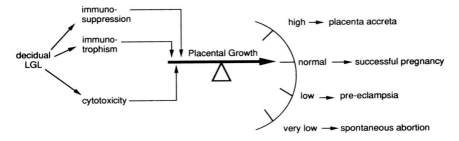

Fig. 7.7 The ways in which decidual LGL might regulate placental (trophoblast) growth. The outcome in placental growth is depicted as a balance between suppression of the maternal immune response and immunotrophism, both of which could promote trophoblast growth, and cytotoxicity or cytostasis which could inhibit trophoblast growth. Normal placental growth would be associated with pregnancy success, low growth with pre-eclampsia, very low growth with spontaneous abortion and high growth with placenta accreta.

4.1 Modulation of the maternal response to placental cells

The pregnant uterus, despite containing large numbers of immune cells, is an immunologically privileged site (56). Animal studies demonstrate that pregnant rodents can reject tissue grafts from semi-allogeneic F1 progeny without harming the embryos *in utero* (57). Part of this immune privilege may be due to the fact that the uterus has relatively poor lymphatic drainage. Lymph vessels are found in the myometrium of the uterus, but in the rodent at least, they are infrequent, small and do not reach the endometrial lumen or the implanting fetus (58). Unlike the eye, where immunological privilege has been suggested to be due to the absence of cells expressing MHC Class II (56), the decidua contains numerous Class II$^+$ macrophages (59). Using unfractionated cell dispersions it has been shown that both mouse and human early pregnancy decidua contain cells which can present antigen both *in vitro* and *in vivo*. The antigen-presenting activity is antigen-specific and MHC-restricted (60–62). In human decidua, the preparations of antigen-presenting cells were 80–90 per cent macrophages, although it was suggested that the sensitivity of the activity to ultraviolet irradiation was more characteristic of dendritic cells (63, 64). In horse pregnancy, the mothers develop circulating cytotoxic antibodies directed against fetal MHC antigens. The stimulus to antibody formation appears to be the invasive trophoblast of the chorionic girdle, which unlike the non-invasive forms of equine trophoblast, express conventional MHC Class II. The speed of the maternal antibody response to this antigen challenge is much faster than the normal primary immune response to an antigen (6), and suggests that in this species antigen-presenting capacity within the decidua is highly efficient.

Decidual cells therefore have the capacity to mount a local immune response to non-self, as both antigen-presenting cells and T-cells are present. It has been suggested that T-cells responding to MHC Class I in the absence of MHC Class II, as would happen with some forms of trophoblast, are more likely to produce a suppressor T-cell response (65). In mouse, T-cells with γ/δ receptors respond preferentially to altered Class I, or antigen associated with Class I, unlike those with α/β receptors which recognize antigen only in association with MHC Class II (66). Most decidual T-cells express the α/β T-cell receptor, but some express the alternative γ/δ form (16). Unlike mouse T-cells, however, in human tissues γ/δ T-cells appear functionally very similar to α/β cells (67).

Given that decidua contains all the cellular elements necessary to mount either an immune or an inflammatory response, attention has focused on the ability of decidual cells or factors to suppress immune function. Unfractionated decidual cells from both mouse and human early pregnancy samples inhibit *in vitro* T-cell proliferation in response to antigen in the mixed lymphocyte reaction (MLR) (68), or in response to mitogen (69, 70). They also inhibit the generation of cytotoxic T-cells in a MLR (71). The effect does not require cell contact, but can be mediated by medium conditioned by incubation with decidual cells (72).

In some instances, all or part of this immunosuppressive activity has been shown to be due to local prostaglandin production (73–75), and the cellular source of these prostaglandins has usually been assumed to be the decidual macrophages (73, 75). However, a major source of prostaglandin synthesis in early decidua is the glandular epithelium where synthesis is inhibited by progesterone (76, 77). A further complicating factor is that decidua appears to contain a soluble factor which competitively inhibits cycloxygenase catalysed prostaglandin synthesis (78).

Extensive investigations of decidual immunosuppressive activity in various mouse matings however, have demonstrated the involvement of cells other than macrophages. Decidual cell dispersions prepared by mechanical sieving were shown to produce a soluble factor which suppressed cytotoxic T-cell generation in a MLR (79). Fractionation of the decidual cells by velocity sedimentation demonstrated that the suppressive activity was associated with cells of small granular lymphocyte morphology (79). Similar immunosuppressive cells had previously been demonstrated in the uterine lymph nodes of pregnant mice soon after mating, with levels peaking at the time of implantation (80). These decidual or lymph node lymphocytes possessed functional Fc receptors (81), but were negative for the T-cell marker, Thy1.2, and for the NK cell marker asialo-GM_1 (82, 83). Selective removal of asialo-GM_1^+ NK cells from decidua taken in the latter half of pregnancy demonstrated that the suppressor cell population was distinct from decidual NK cells (81). Mouse metrial gland cells also produce soluble factors with immunosuppressive activity *in vitro* (84). The suppressive factor from mouse decidua was originally shown to have a molecular weight of less than 100 000 kDa and to inhibit the IL-2-dependent proliferation of T-cells. The effect is not overcome by addition of exogenous IL-2 (82), and does not involve any reduction in IL-2 receptor number (72).

More recently, a suppressor factor has been purified by HPLC and shown to have the biological characteristics of transforming growth factor β (TGFβ). Specific antibodies identify it as being closely related to TGFβ$_2$. A similar molecule has been identified in human decidua, and *in situ* hybridization with a cDNA probe for TGFβ$_2$ demonstrated small numbers of positive cells in human first trimester decidua (85, 86).

The identity of the suppressor cells described by Clark and co-workers is still not clear. Two types are mentioned which can be distinguished on the basis of cell size (83). The larger cells are present early in mouse pregnancy, and analogous cells were also seen (69, 70) in the secretory phase of the menstrual cycle and in early pregnancy in women, when CD56$^+$ LGLs are most abundant. Human decidual LGL clones have been shown to secrete active TGFβ (16). Most characterization, however, has been done with the smaller cells in mouse which are present throughout pregnancy. These do not share the surface characteristics of either T-cells or NK cells, nor the NK activity of LGLs. They are present throughout pregnancy, which is usually 21 days in the mouse, but most commonly have been isolated between days 12 and 18. They express Fc receptors, and the possibility that they are macrophages has not been conclusively excluded.

The role of these suppressor cells in successful pregnancy has been investigated in a number of mouse matings which are associated with a high level of spontaneous abortion. In *Mus musculus* mothers carrying *Mus caroli* embryos or hybrid embryos with *Mus musculus* inner cell mass surrounded by *Mus caroli* trophoblast, the influx of suppressor cells which occurs in normal pregnancy in the uterine lymph nodes and decidua does not happen. Instead, the decidua becomes infiltrated by cytotoxic T-cells and the embryos are resorbed. It has been suggested that the failure of *Mus caroli* trophoblast to recruit maternal suppressor cells, allows rejection of the embryo by the maternal immune system, and that this is a model for spontaneous abortion (71, 87). Subsequent studies however have demonstrated that embryo death still occurs in immunocompromised mothers, and suggest that the defect is simply the inability of the *Mus caroli* trophoblast to develop in *Mus musculus* (88).

An alternative model has been developed from studies of the abortionprone mating of CBA/J and DBA/2J mouse strains, where the resorption rate is increased by presensitization of the mother with paternal cells, and is associated with a deficiency of decidual suppressor cells (83). In contrast to the effect of pre-immunization with DBA/2 cells, immunization with cells from BALB/c mice, which when mated with CBA/J females do not have a high spontaneous resorption rate, reduces the resorption rate in subsequent CBA/J versus DBA/2 matings (89). Using a variety of recombinant cell lines created between DBA/2 and BALB/c, to pre-immunize before CBA/J versus DBA/2 matings, it has been shown that cell lines which increase the resorption rate decrease decidual suppressor cell activity, and vice versa (90). Others have demonstrated that the conditions which cause decreased suppressor cell activity are associated with increased decidual NK cell activity (91).

In humans, results from histological studies of decidua taken from women who had spontaneously aborted claimed to show that granulated cells, presumed to correspond to decidual LGLs, were much less numerous in these women than in normal pregnancies (92).

To summarize, there is clear evidence from the mouse that decidual cells produce an immunosuppressive factor which is related to $TGF\beta_2$. The cells responsible for this production seem to be small granular lymphocytes, and the absence of such cells from the decidua is associated with an increased risk of spontaneous abortion. A similar immunosuppressive factor is produced by human decidua. The cells responsible have not been positively identified, though cloned decidual LGLs, at least after activation with IL-2 and mitogens, have the ability to produce $TGF\beta$. Spontaneous abortion in women may also be associated with a reduction in decidual LGLs.

The immunosuppressive factors described in pregnancy decidua are presumed to inhibit the response the decidual T-cells would otherwise make to the invading trophoblast. Certainly their immunosuppressive activity has always been measured against T-cells. As mentioned in the introduction to this chapter, there are doubts as to whether T-cells would normally mount a rejection response against human trophoblast; mouse trophoblast is different in expressing conventional MHC Class I antigens. Nevertheless, there are circumstances, such as bacterial infections or inflammation, where decidual T-cells might become activated and release such cytokines as IL-2 and IL-4.

Immunohistology and flow cytometry of isolated cells has demonstrated that decidual $CD56^+$ LGLs express receptors for both IL-2 and IL-4. LGLs are the only cells in decidua to express the p75 intermediate affinity IL-2R protein, but do not express the p55 (CD25) receptor protein. Decidual T-cells appear to express neither CD25 (93) nor the p75 protein. Using an antibody directed against the IL-4 receptor (94), some decidual LGLs were shown to be $IL-4R^+$, as were some decidual macrophages and most T-cells (Fig. 7.8). The IL-4R was also expressed on the basal face of the glandular epithelium (134). Decidual LGLs have been shown to proliferate in response to IL-2, which also increases their cytotoxic activity (see below).

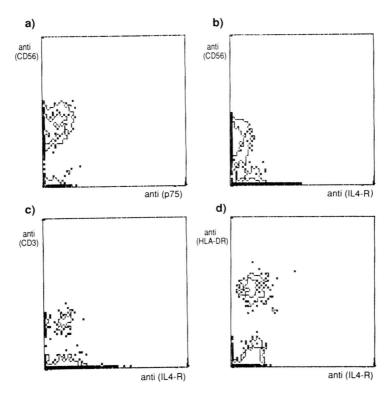

Fig. 7.8 Flow cytometric analysis of the expression of the p75 IL-2 receptor and the IL-4 receptor on decidual LGLs, T-cells, and macrophages. Cell dispersions from first trimester decidua were double labelled with one monoclonal antibody being detected with fluoresceinated second antibody, and the other monoclonal being directly phycoerythrinated, or biotinylated and detected with phycoerythrin-streptavidin. The samples were analysed by flow cytometry and the profiles show log peak green fluorescence on the horizontal axis and log peak red fluorescence on the vertical axis. Antibodies used were: (a) anti-p75 (132) and anti-CD56; (b) anti-IL-4 receptor (133) and anti-CD56; (c) anti-IL-4R and anti-CD3; and (d) anti-IL-4R and anti-HLA-DR.

IL-4 alone was found to have no effect on decidual LGLs, but it did inhibit the proliferation of LGL induced by IL-2 (Fig. 7.9). IL-2 did not up-regulate either the p75 protein or the IL-4R, nor did it induce expression on decidual LGL of CD25 (134).

In their expression of IL-2R proteins, and their proliferative response to IL-2 and IL-4, the decidual LGL resemble the CD16^{dim}/CD56^{bright} subset of blood LGLs, rather than the CD16$^-$/CD56^{bright} subset (15).

IL-4 in many situations, plays an immunosupressive role. It is not known

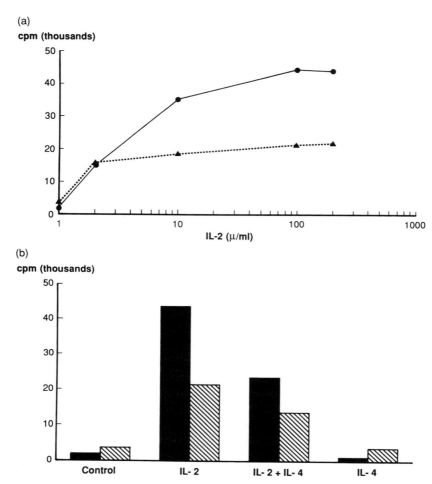

Fig. 7.9 Proliferative response of CD56[+] decidual LGLs to IL-2 and IL-4. (a) Two preparations of CD56[+] decidual LGL were purified by flow cytometry to 96% (●) or 98% (▲) and incubated with recombinant IL-2 for 3 days. Proliferation was measured as incorporation of ^{3}H-thymidine during a 24 h pulse. (b) The same two preparations of CD56[+] cells as in fig. 9a were incubated as before with recombinant IL-2 (200 U/ml) and IL-4 (400 U/ml) either separately or together (opaque bars, 96% CD56[+] cell population; hatched bars, 98% CD56[+] cell population).

whether IL-4 is produced in pregnancy decidua, although both T-cells and decidual LGLs might be expected to be a source of the cytokine (95). Unstimulated decidual LGLs produce no detectable IL-2 (96). It is clear, however, that were IL-4 to be produced in decidua it could limit the effect of IL-2 in activating decidual LGLs. IL-4 also has the potential for modulating the behaviour of other immune cells within the decidua which express the IL-4R: macrophages, T-cells, and the glandular epithelium. The latter is itself known to be a source of other cytokines including M-CSF (97) and IFNα and γ (98).

4.2 Inhibition of trophoblast development

The ability of the decidua to inhibit trophoblast invasiveness is suggested by the behaviour of trophoblast in ectopic pregnancies. In the pig, trophoblast *in utero* does not invade the uterine epithelium, nor form a syncytium, but in contrast, ectopic porcine trophoblast is extremely invasive and does differentiate to a syncytium (99). Similarly in the mouse, trophoblast is much more invasive and destructive of maternal tissue ectopically than *in utero* (100). In human pregnancy, immunohistology of first trimester tissue with antibody to the proliferation marker, Ki67, indicates that unlike the cytotrophoblast columns of the anchoring villi, and villous cytotrophoblast, extra-villous cytotrophoblast within the decidua are Ki67-negative (37), suggesting that the decidua might have an inhibitory effect on trophoblast proliferation.

The most obvious way in which decidua might limit trophoblast invasiveness would be by cell-mediated cytotoxicity. Unfractionated decidual cells from mouse (101) and human (96, 102–104) decidua have been shown to have cytotoxic activity as measured against conventional NK cell tumour targets. In the mouse, cells with NK activity have been indirectly identified as positive for asialo-GM_1 a marker of NK cells, and are present in largest numbers at day 6 of pregnancy (101). They appear to be distinct from the asialo-GM_1^- suppressor cells which become more numerous later on (81), and from the murine metrial gland cells, which also resemble LGLs, but have no cytotoxic activity against NK cell targets (84). In the mouse matings with a high degree of spontaneous abortion used by others to study decidual immunosuppressive factors (see Section 4.1), NK cells identified as asialo-GM_1^+ cells, were detected in the decidua near resorbing fetuses. Pre-immunization with cells from non-abortion prone strains decreased the levels of asialo-GM_1^+ cells and the abortion rate (91). Depletion of NK cells with antibody to asialo-GM_1 reduced the resorption rate, and injection with polyinosinic/cytidylic acid which activates NK cells, increased it (105).

In human decidua, CD56$^+$ LGLs have been isolated by flow cytometry to more than 98 per cent purity and shown to be the cell population mediating the NK activity of first trimester decidua (96) (Fig. 7.10). Their NK activity is, however, low compared with that of peripheral blood cells and cannot be increased by preincubation of decidual cells with IFNγ (104). This is not unexpected, as the CD56^{bright} subsets of blood LGLs have low cytotoxic activity compared with the CD16^{bright}/CD56^{dim} population (15, 106).

The demonstration that decidual cells have NK activity is only relevant

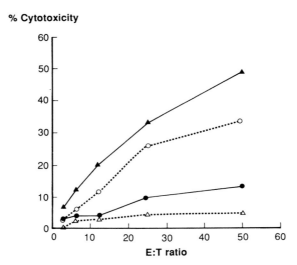

% Cytotoxicity

Fig. 7.10   Cytotoxic activity against K562 target cells, of decidual LGLs. Effector cells were prepared from cell dispersions of first trimester decidua, as either unfractionated decidual cells (●), flow cytometrically purified CD56$^+$ LGLs (○), or CD56$^-$ decidual cells, a mixture of macrophages, T-cells and stromal cells, (△). Peripheral blood mononuclear cells were included as a control (▲). Cytotoxicity was measured in 6 h chromium release assays against K562 targets. Values are the medians from 12 experiments. (Data are taken from (96).)

to the control of placentation if they could be demonstrated to have cytotoxic or cytostatic activity against trophoblast. Neither unfractionated human decidual cells (102) nor purified CD56$^+$ decidual LGLs (107) showed any cytotoxic activity against freshly isolated human trophoblast. In the mouse system, there is some evidence that NK cells bind to trophoblast without killing (108). Indeed, mouse trophoblast seem very resistant to cell-mediated cytotoxicity (109). Thus, two-day cultured mouse trophoblast, despite being bound by cytotoxic T-cells and spleen NK cells, were not killed by either, even after pretreatment with IFN (110, 111). Some mouse trophoblast, unlike human cells, express conventional MHC Class I antigens. If coated with antibodies to Class I, the murine trophoblast could not be lysed by NK cells, demonstrating their resistance to antibody dependent cell-mediated cytotoxicity, but were lysed by antibody and complement (111). Cultured mouse trophoblast could be killed by cells with very high cytotoxic activity, such as spleen LAK cells or cytotoxic T-cells generated *in vitro* in special medium, but fresh trophoblast remained resistant to killing (112, 113).

Human decidual LGLs incubated *in vitro* with IL-2 showed increased cytotoxic activity against NK target cell lines (96, 103), and acquired low

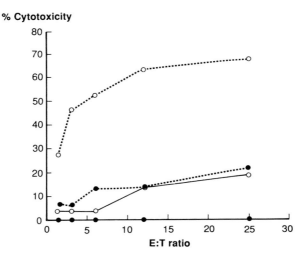

Fig. 7.11 Cytotoxic activity of decidual cells before and after incubation with IL-2. Effector cells were unfractionated cell dispersions prepared from first trimester decidua, freshly isolated (——) or incubated for 7 days in 200 U/ml recombinant IL-2 (·········). Target cells were K562 (○) or the choriocarcinoma cell line, BeWo (●). Values are the medians from 6 experiments. (Data taken from (107).)

cytotoxicity against BeWo (Fig. 7.11) and JEG choriocarcinoma cells and fetal fibroblasts (107, 135). Even these activated LGLs were unable to kill freshly isolated term human trophoblast which seemed to have no surface NK target structures (107), but activated LGLs did have low cytotoxic activity against trophoblast cultured from first trimester tissue (115). It has been suggested from experiments using decidual LGL clones (16) that the increased cytotoxicity induced by IL-2 is due to the preferential expansion of the more cytotoxic CD16^bright/CD56^dim population. However, when unfractionated decidual cells were incubated with IL-2 under conditions which led to increased cytotoxic activity, the activated LGLs were still CD16^−/CD56^bright (96).

It appears therefore that trophoblast are likely in most circumstances *in vivo* to be resistant to cell-mediated cytotoxicity either by decidual LGLs or infiltrating cytotoxic lymphocytes or LAK cells. The possibility remains that decidual LGLs might have cytostatic activity against trophoblast. Using rat trophoblast-derived cell lines, Hunt and co-workers (114) demonstrated that, of the variety of cytokines tested, TNFα and TGFβ were both cytostatic for trophoblast cell lines, while having no effect on rat fibroblasts. In a different study, TNFα was shown not to be cytotoxic for mouse trophoblast (115), although as these cells did not proliferate *in vitro*, cytostatic effects would not have been detected.

TGFβ is present in mouse and human decidua, and the evidence that decidual LGLs may be a source has been discussed above. TNFα has also been shown to be synthesized by human decidual cells from tissue taken in the first trimester (116) and at term (117). Some cloned decidual LGLs, when cultured in the presence of phytohaemagglutanin (PHA), produced low levels of TNFα (16). However, using decidual macrophages purified to 95 per cent purity by flow cytometry, we have demonstrated that for unstimulated cells both TNFα and TNFβ mRNAs and biologically active TNF are localized exclusively to decidual macrophages (116). TNFα mRNA was not detected in RNA isolated from purified decidual LGLs (G. Vince, unpublished observations).

4.3   Regulation of trophoblast growth

In contrast to a possible role in limiting trophoblast growth, there is also some evidence that the decidua may produce factors which positively promote placental development. Wegmann and co-workers, from studies of mouse pregnancy, have developed the theory of 'immunotrophism'. They noted that in their model system for spontaneous abortion, immunization of female mice with paternal spleen cells prior to conception, as well as reducing the rate of abortion, increased both fetal and placental weight in the surviving fetuses. Placental weight was also found to be greater in F1 hybrid pregnancies than in inbred matings. If pregnant mice were depleted of T-cells by treatment with anti-T-cell antibody between days 8 and 10 of gestation, then placental weight was decreased (118, 119). They suggested that recognition of the fetal allograft by the maternal immune cells resulted not only in immunosuppression of the maternal rejection response, but also in the production of factors by the immune cells which stimulated placental growth. They suggested that the CSFs might be the factors involved, since the proliferation of unfractionated murine placental cells *in vitro* was found to be stimulated by the addition of recombinant human M-CSF, GM-CSF or IL-3 (120). Mouse and human trophoblast and human choriocarcinoma cells have been shown to express the receptor for M-CSF (121–123) and the receptor for G-CSF has been detected on human placental membranes and on choriocarcinomas (124).

The source of CSFs was presumed by Wegmann and colleagues to be the maternal immune cells since *in vivo* treatment of the mother with antibodies which depleted circulating thymocytes or T-cells reduced *in vivo* incorporation of ^{3}H-thymidine in the placenta (120). By analogy with the equivalent blood populations both the small numbers of decidual T-cells and decidual LGLs might be expected to produce CSFs. Blood LGLs release GM-CSF, M-CSF, and IL-3 in response to a variety of stimuli, although G-CSF was not detected, and unstimulated LGLs contained no CSF mRNA (125).

However, the only hard data indicate that in the mouse the major source of M-CSF in the uterus, where levels increase a thousandfold in pregnancy, is the endometrial glands (126). Synthesis of M-CSF by the glandular epithelium is increased by progesterone, or indirectly by human chorionic gonadotrophin (97). M-CSF synthesis in human decidua has not been investigated, but decidual cells taken from first trimester tissue, have been shown to synthesize biologically active G-CSF, and mRNA for G-CSF has been localized to decidual macrophages (136).

5 Summary

The LGL population found in such abundance in the human decidua during early pregnancy is unusual in a number of ways. Most decidual LGLs, in their antigen expression and response to cytokines, appear to correspond to the CD16^{dim}/CD56^{bright} minor subset of blood LGLs, and no similar population has been reported in tissues other than decidua. The recruitment of LGLs to the non-pregnant endometrium and to the decidua in early pregnancy, and/or their proliferation *in situ*, appears to be under hormonal control, though it is not clear whether this is due to a direct effect of the sex steroid hormones on LGLs themselves, or is mediated by another hormone-responsive cell type.

The abundance of decidual LGLs in early pregnancy, their virtual disappearance as pregnancy progresses and their unusual phenotype, all suggest that they must play a vital role in early pregnancy, possibly contributing to success of the pregnancy by regulating placental development.

Although our knowledge of the properties of decidual LGLs does provide some evidence of their role in decidua, the picture is still only fragmentary. The evidence from the mouse system that at least some decidual LGLs may produce TGFβ, which by suppressing the maternal immune system in the decidua may permit trophoblast invasion, is quite strong. It is also clear that if a local immune or inflammatory response led to the production of IL-2 and IL-4, decidual LGLs would respond to IL-2 with proliferation and increased cytotoxic activity, but that IL-4 would inhibit at least their proliferative activity. This suggests a second mechanism, involving IL-4, for local immunosuppression in the decidua. However, evidence for a direct effect of decidual LGLs on trophoblast growth and development is more tenuous. Although decidual LGLs have cytotoxic activity, it is weak, and even after activation with IL-2, killing of trophoblast is rare. It is possible that decidua may produce a factor which enhances the killing efficacy of decidual LGLs. TNFα, for example, which is produced by decidual macrophages (116), as well as having a direct cytostatic effect on rat trophoblast (114), is known to increase LGL cytotoxic activity, both alone and in combination with IL-2 (127). It is possible that if TNFα and

IL-2 had a comparable synergistic effect on the cytotoxic activity of decidual LGLs, then killing of trophoblast might occur. Circumstantial evidence in both mouse and human systems certainly suggests that increased LGL cytotoxic activity may be involved in spontaneous abortion. No direct evidence has yet been found of a role for decidual LGLs in stimulating trophoblast or placental growth.

Our understanding of how decidual LGLs interact with all the other cell types present in early pregnancy decidua is even more rudimentary. What is now clear is that the decidua in pregnancy is a highly complex tissue containing not only immune cell populations, but also specialized stromal cells and glandular epithelium. All of these cells have the potential to secrete a variety of cytokines: endometrial stromal cells secrete IL-6 in response to IL-1 or TNFα (128); glandular epithelial cells in mouse secrete M-CSF (97); and in human may synthesize IFNα and γ (98). Decidual macrophages secrete TNFα (116) and G-CSF (136). Decidual LGLs are part of this complex network of cell- and cytokine-mediated interactions, and we are only starting to unearth the role of this unique cell population in the success and failure of pregnancy.

References

1 Fox, H. and Jones, C. J. P. (1983). Pathology of the trophoblast. In *Biology of the trophoblast* (ed. Y. W. Loke and A. Whyte), pp. 137–85. Elsevier, Amsterdam.

2 Sunderland, C. A., Naiem, M., Mason, D. Y., Redman, C. W. G., and Stirrat, G. M. (1981). The expression of major histocompatibility antigens by human chorionic villi. *J. Reprod. Immunol.*, **3**, 323–31.

3 Ellis, S. A., Palmer, M. S., and McMichael, A. J. (1990). Human trophoblast and the choriocarcinoma cell line BeWo express a truncated HLA Class I molecule. *J. Immunol.*, **144**, 731–5.

4 Starkey, P. M. (1987). Reactivity of human trophoblast with an antibody to the HLA class II antigen, HLA-DP. *J. Reprod. Immunol.*, **11**, 63–70.

5 Billington, W. D. and Bell, S. c. (1983). Immunobiology of the mouse trophoblast. In *Biology of the trophoblast* (ed. Y. W. Loke and A. Whyte), pp. 571–95. Elsevier, Amsterdam.

6 Antczak, D. F. and Allen, W. R. (1988). A non-genetic development defect in trophoblast formation in the horse: immunological aspects of a model of early abortion. In *Early pregnancy loss* (ed. R. W. Beard and F. Sharp), pp. 123–39. Royal College of Obstetricians and Gynaecologists, London.

7 Bulmer, J. N., Wells, M., Bhabra, K., and Johnson, P. M. (1986). Immunohistological characterization of endometrial gland epithelium

and extravillous foetal trophoblast in third trimester human placental bed tissues. *Br. J. Obstet. Gynaecol.*, **93**, 823–32.

8 Bulmer, J. N, and Sunderland, C. A. (1983). Bone-marrow origin of endometrial granulocytes in the early human placental bed. *J. Reprod. Immunol.*, **5**, 383–7.

9 Ritson, A. and Bulmer, J. N. (1987). Endometrial granulocytes in human decidua react with a natural-killer (NK) cell marker, NKH1. *Immunology*, **62**, 329–31.

10 Bulmer, J. N., Hollings, D., and Ritson, A. (1987). Immunocyto-chemical evidence that endometrial stromal granulocytes are granu-lated lymphocytes. *J. Pathol.*, **153**, 281–88.

11 King, A., Wellings, V., Gardner, L., and Loke, Y. W. (1989). Im-munocytochemical characterisation of the unusual large granular lym-phocytes in human endometrium throughout the menstrual cycle. *Human Immunol.*, **24**, 195–205.

12 Hamperl, H. and Hellweg, G. (1958). Granular endometrial stroma cells. *Obstet. Gynecol.*, **11**, 379–87.

13 Bulmer, J. N. and Johnson, P. M. (1984). Macrophage populations in the human placenta and amniochorion. *Clin Exp. Immunol.*, **57**, 393–403.

14 Starkey, P. M., Sargent, I. L., and Redman, C. W. G. (1988). Cell populations in human early pregnancy decidua: characterization and isolation of large granular lymphocytes by flow cytometry. *Immuno-logy*, **65**, 129–34.

15 Nagler, A., Lanier, L. L., Cwirla, S., and Phillips, J. H. (1989). Comparative studies of human FcRIII-positive and -negative natural killer cells. *J. Immunol.*, **143**, 3183–91.

16 Christmas, S. E., Bulmer, J. N., Meager, A., and Johnson, P. M. (1990). Phenotypic and functional analysis of human CD3-decidual leucocyte clones. *Immunology*, **71**, 182–9.

17 Bernard, O., Scheid, M. P., Ripoche, M.-A., and Bennett, D. (1978). Immunological studies of mouse decidual cells. I. Membrane markers of decidual cells in the days after implantation. *J. Exp. Med.*, **148**, 580–91.

18 Kearns, M. and Lala, P. K. (1985). Characterization of hemotogenous cellular constituents of the murine decidua: a surface marker study. *J. Reprod. Immunol.*, **8**, 213–34.

19 Kearns, M. and Lala, P. K. (1985). Radioautoradiographic analysis of surface markers on decidual cells shared by cells of the lympho-myeloid tissues. *Am. J. Reprod. Immunol. Microbiol.*, **9**, 39–47.

20 Bulmer, D., Peel, S., and Bulmer, J. N. (1985). The function of the metrial gland in pregnancy. *Res. Reprod.*, **17**, 1–2.

21 Peel, S., Stewart, I., and Bulmer, D. (1982). The morphology of granulated metrial gland cells in lethally-irradiated, bone marrow-reconstituted mice. *J. Anat.*, **135**, 849.

22 Lee, C. S., Gogolin-Ewens, K., and Brandon, M. R. (1988). Identification of a unique lymphocyte subpopulation in the sheep uterus. *Immunology,* **63,** 157–64.

23 Starkey, P. M., Clover, L. M., and Rees, M. C. P. (1991). Variation during the menstrual cycle of immune cell populations in human endometrium. *Eur. J. Obstet. Gynecol.,* **39,** 203–7.

24 Kabawat, S. E., Mostoufi-Zadeh, M., Driscoll, S. G., and Bhan, A. K. (1985). Implantation site in normal pregnancy. A study with monoclonal antibodies. *Am. J. Pathol.,* **118,** 76–84.

25 Khong, T. Y. (1987). Immunohistologic study of the leukocytic infiltrate in maternal uterine tissues in normal and pre-eclamptic pregnancies at term. *Am. J. Reprod. Immunol. Microbiol.,* **15,** 1–8.

26 Vince, G. S., Starkey, P. M., Jackson, M. C., Sargent, I. L., and Redman, C. W. G. (1990). Flow cytometric characterisation of cell populations in human pregnancy decidua and isolation of decidual macrophages. *J. Immunol. Meth.,* **132,** 181–9.

27 Van Vorrhis, W. C., *et al.* (1983). Specific anti-mononuclear phagocyte monoclonal antibodies. Application to the purification of dendritic cells and the tissue localization of macrophages. *J. Exp. Med.,* **158,** 126–45.

28 Bell, S. C. (1983). Decidualization and associated cell types: implications for the role of the placental bed in the materno-foetal immunological relationship. *J. Reprod. Immunol.,* **5,** 185–94.

29 Earl, U., Lunny, D. P., and Bulmer, J. N. (1987). Leucocyte populations in ectopic tubal pregnancy. *J. Clin. Pathol.,* **40,** 901–5.

30 Abel, M. H., Smith, S. K., and Baird, D. T. (1980). Suppression of concentration of endometrial prostaglandin in early intra-uterine and ectopic pregnancy in women. *J. Endocr.,* **85,** 379–86.

31 Dallenbach-Hellweg, G. (1981). *Histopathology of the endometrium.* Springer, Berlin.

32 Pauerstein, C. J., Croxatto, H. B., Eddy, C. A., Ramzy, I., and Walters, M. D. (1986). Anatomy and pathology of tubal pregnancy. *Obstet. Gynecol.,* **67,** 301–8.

33 Tabibzadeh, S. S. and Satyaswaroop, P. G. (1989). Sex steroid receptors in lymphoid cells of human endometrium. *Am. J. Clin. Pathol.,* **91,** 656–63.

34 Gerdes, J., Lemke, H., Baisch, H., Wacker, H. H., Schwab, U., and Stein, H. (1984). Cell cycle analysis of a cell proliferation-associated human nuclear antigen defined by the monoclonal antibody, Ki67. *J. Immunol.,* **133,** 1710–15.

35 Pace, D., Morrison, L., and Bulmer, J. N. (1989). Proliferative activity in endometrial stromal granulocytes throughout menstrual cycle and early pregnancy. *J. Clin. Pathol.,* **42,** 35–9.

36 Tabibzadeh, S. (1990). Proliferative activity of lymphoid cells in human endometrium throughout the menstrual cycle. *J. Clin. Endocrinol. Metab.*, **70**, 437–43.

37 Bulmer, J. N., Morrison, L., and Johnson, P. M. (1988). Expression of the proliferation markers Ki67 and transferrin receptor by human trophoblast populations. *J. Reprod. Immunol.*, **14**, 291–302.

38 Krcek, J. and Clark, D. A. (1985). Selective localization of a bone marrow cell subpopulation at the implantation site in murine decidua. *Am. J. Reprod. Immunol. Microbiol.*, **7**, 95–8.

39 Chatterjee-Hasrouni, S., Santer, V., and Lala, P. K. (1980). Characterisation of maternal small lymphocyte subsets during allogeneic pregnancy in the mouse. *Cell. Immunol.*, **50**, 290–304.

40 Lanier, L. L., Le, A. M., Civin, C. I., Loken, M. R., and Phillips, J. H. (1986). The relationship of CD16 (Leu-11) and Leu-19 (NKH-1) antigen expression on human peripheral blood NK cells and cytotoxic T lymphocytes. *J. Immunol.*, **136**, 4480–6.

41 Lanier, L. L., Testi, R., Bindl, J., and Phillips, J. H. (1989). Identity of Leu-19(CD56) leukocyte differentiation antigen and neural cell adhesion molecule. *J. Exp. Med.*, **169**, 2233–8.

42 Nitta, T., Yagita, H., Sato, K., and Okumura, K. (1989). Involvement of CD56 (NKH-1/Leu 19 antigen) as an adhesion molecule in natural killer-target cell interaction. *J. Exp. Med.*, **170**, 1757–61.

43 Jalkanen, S. T., Bargatze, R. F., Herron, L. R., and Butcher, E. C. (1986). A lymphoid cell surface glycoprotein involved in endothelial cell recognition and lymphocyte homing in man. *Eur. J. Immunol.*, **16**, 1195–1202.

44 Miller, L. J., Bainton, D. F., Borregaard, N., and Springer, T. A. (1987). Stimulated mobilization of monocyte Mac-1 and p150, 95 adhesion proteins from an intracellular vesicular compartment to the cell surface. *J. Clin. Invest.*, **80**, 535–44.

45 Sulke, A. N., Jones, D. B., and Wood, P. J. (1985). Variation in natural killer activity in peripheral blood during the menstrual cycle. *Br. Med. Bull.*, **290**, 884–6.

46 Gregory, C. D., Shah, L. P., Lee, H., Scott, I. V., and Golding, P. R. (1985). Cytotoxic reactivity of human natural killer (NK) cells during normal pregnancy: a longitudinal study. *J. Clin. Lab. Immunol.*, **18**, 175–82.

47 Asari, S., Iwatani, Y., Anuno, N., Tanizawa, O., and Mujai, K. (1989). Peripheral K cells in normal human pregnancy decrease during pregnancy and increase after delivery. *J. Reprod. Immunol.*, **15**, 31–7.

48 Gregory, C. D., Lee, H., Scott, I. V., and Golding, P. R. (1987). Phenotypic heterogeneity and recycling capacity of natural killer cells in normal human pregnancy. *J. Reprod. Immunol.*, **11**, 135–45.

49 Legendre, C. M., Guttmann, R. D., and Yip, G. H. (1986). Natural

killer cell subsets in long-term renal allograft recipients. A phenotypic and functional study. *Transplantation*, **42**, 347–52.

50 Toder, V., Nebel, L., and Gleicher, N. (1984). Studies of natural killer cells in pregnancy. I. Analysis at the single cell level. *J. Clin. Lab. Immunol.*, **14**, 123–7.

51 Toder, V., Nebel, L., Elrad, H., Blank, M., Durdana, A., and Gleicher, N. (1984). Studies of natural killer cells in pregnancy. II. The immunoregulatory effect of pregnancy substances. *J. Clin. Lab. Immunol.*, **14**, 129–33.

52 Ahmed, S. A., Penhale, W. J., and Talal, N. (1985). Sex hormones, immune responses, and autoimmune diseases. Mechanisms of sex hormone action. *Am. J. Pathol.*, **121**, 531–51.

53 Szekeres-Bartho, J., Hadnagy, J., Csernus, V., Balazs, L., Magyarlaki, T., and Pacso, A. S. (1985). Increased NK activity is responsible for higher cytotoxicity to HEF cells by lymphocytes of women with threatened pre-term delivery. *Am. J. Reprod. Immunol. Microbiol.*, **7**, 22–6.

54 Toder, V., Blank, M., Gleicher, N., Voljovich, I., Mashiah, S., and Nebel, L. (1983). Activity of natural killer cells in normal pregnancy and edema-proteinuria-hypertension gestosis. *Am. J. Obstet. Gynecol.*, **145**, 7–10.

55 Alanen, A. and Lassila, O. (1982). Deficient natural killer cell function in pre-eclampsia. *Obstet. Gynaecol.*, **60**, 631–4.

56 Streilein, J. W. and Wegmann, T. G. (1987). Immunologic privilege in the eye and the foetus. *Immunol. Today*, **8**, 362–6.

57 Woodruff, M. F. A. (1958). Transplantation immunity and the immunological problem of pregnancy. *Proc. Roy. Soc. London. (Ser. B)*, **148**, 68–75.

58 Head, J. R. (1987). Lymphoid components in the rodent uterus. In *Immunoregulation and foetal survival* (ed. T. J. Gill and T. G. Wegmann), pp. 46–59. Oxford University Press.

59 Bulmer, J. N. and Johnson, P. M. (1985). Immunohistological characterisation of the decidual leucocytic infiltrate related to endometrial gland epithelium in early human pregnancy. *Immunology*, **55**, 35–44.

60 Elcock, J. M. and Searle, R. F. (1985). Antigen-presenting capacity of mouse decidual tissue and placenta. *Am. J. Reprod. Immunol.*, **7**, 99–103.

61 Dorman, P. J. and Searle, R. F. (1988). Alloantigen presenting capacity of human decidual tissue. *J. Reprod. Immunol.*, **13**, 101–12.

62 Matthews, C. J. and Searle, R. F. (1988). Antigen-presenting capacity of murine decidual tissue *in vivo J. Reprod. Immunol.*, **12**, 287–95.

63 Oksenberg, J. R., Mor-Yosef, S., Persitz, E., Schenker, J., Mozes, E., and Brautbar, C. (1986). Antigen-presenting cells in human decidual tissue. *Am. J. Reprod. Immunol. Microbiol.*, **11**, 82–8.

64 Oksenberg, J. R., Mor-Yosef, S., Ezra, Y., and Brautbar, C. (1987). Antigen presenting cells in human decidual tissue. II. accessory cells for the development of anti-trinitrophenyl cytotoxic T lymphocytes. *J. Reprod. Immunol.*, **10**, 309–18.

65 Goeken, N. E. (1984). Human suppressor cell induction *in vitro*: preferential activation by Class I MHC antigen. *J. Immunol.*, **132**, 2291–9.

66 Janeway, C. A. Jr., Jones, B., and Hayday, A. (1988). Specificity and function of T cells bearing τδ receptors. *Immunol. Today*, **9**, 73–6.

67 Kozbor, D., *et al.* (1989). Human TCR-τ+/δ+, CD8⁺ T lymphocytes recognize tetanus toxoid in an MHC-restricted fashion. *J. Exp. Med.*, **169**, 1847–51.

68 Nakayama, E., Asano, S., Kodo, H., and Miwa, S. (1985). Suppression of mixed lymphocyte reaction by cells of human first trimester pregnancy endometrium. *J. Reprod. Immunol.*, **8**, 25–31.

69 Daya, S., Clark, D. A., Devlin, C., Jarrell, J., and Chaput, A. (1985). Preliminary characterisation of two types of suppressor cells in the human uterus. *Fertil. Steril.*, **44**, 778–85.

70 Daya, S., Clark, D. A., Devlin, C., Jarrell, J., and Chaput, A. (1985). Suppressor cells in human decidua. *Am. J. Obstet. Gynecol.*, **151**, 267–70.

71 Clark, D. A., Slapsys, R., Croy, B. A., Krcek, J., and Rossant, J. (1984). Local active suppression by suppressor cells in the decidua: a review. *Am. J. Reprod. Immunol.*, **5**, 78–83.

72 Clark, D. A., Falbo, M., Rowley, R. B., Banwatt, D., and Stedronska-Clark, J. (1988). Active suppression of host-vs-graft reaction in pregnant mice. IX. Soluble suppressor activity obtained from allopregnant mouse decidua that blocks the cytolytic effector response to IL-2 is related to transforming growth factor-β. *J. Immunol.*, **141**, 3833–40.

73 Lala, P. K., Parhar, R. S., Kearns, M., Johnson, S., and Scodras, J. M. (1986). *Immunological aspects of the decidual response*, Proceedings of the 3rd International Congress on Reproduction and Immunology (ed. D. A. Clark and B. A. Croy), pp. 190–8. Elsevier, Amsterdam.

74 Matthews, C. J. and Searle, R. F. (1987). The role of prostaglandins in the immunosuppressive effects of supernatants from adherent cells of murine decidual tissue. *J. Reprod. Immunol.*, **12**, 109–24.

75 Wood, G. W., Kamel, S., and Smith, K. (1988). Immunoregulation and prostaglandin production by mechanically-derived and enzyme-derived murine decidual cells. *J. Reprod. Immunol.*, **13**, 235–48.

76 Smith, S. K. and Kelly, R. W. (1987). The effect of estradiol-17β and actinomycin D on the release of PGF and PGE from separated cells of human endometrium. *Prostaglandins*, **34**, 553–61.

77 Smith, S. K. and Kelly, R. W. (1987). The effect of the antiprogestins

RU 486 and ZK 98734 on the synthesis and metabolism of prostaglandins F-2α and E-2 in separated cells from early human decidua. *J. Clin. Endocrinol. Metab.*, **65**, 527–34.

78 Ishihara, O., Kinoshita, K., Satoh, K., and Mizuno, M. (1987). The inhibitory effect of cytosolic fraction of human decidua on prostaglandin synthesis. *Endocrinol. Japon.*, **34**, 793–8.

79 Slapsys, R. M. and Clark, D. A. (1982). Active suppression of host-vs-graft reaction in pregnant mice. IV. Local suppressor cells in decidua and uterine blood. *J. Reprod. Immunol.*, **4**, 355–64.

80 Clark, D. A. and McDermott, M. R. (1981). Active suppression of host-vs-graft reaction in pregnant mice. III. Developmental kinetics, properties and mechanism of induction of suppressor cells during first pregnancy. *J. Immunol.*, **127**, 1267–73.

81 Slapsys, R. M., Richards, C. D., and Clark, D. A. (1986). Active suppression of host-versus-graft reaction in pregnant mice. VIII. The uterine decidua-associated suppressor cell is distinct from decidual NK cells. *Cell. Immunol.*, **99**, 140–9.

82 Clark, D. A., Chaput, A., Walker, C., and Rosenthal, K. L. (1985). Active suppression of host-vs-graft reaction in pregnant mice. VI. Soluble suppressor activity obtained from decidua of allopregnant mice blocks the response to IL2. *J. Immunol.*, **134**, 1659–64.

83 Clark, D. A., *et al.* (1986). Immunoregulatory molecules of trophoblast and decidual suppressor cell origin at the maternofoetal interface. *Am. J. Reprod. Immuol. Microbiol.*, **10**, 100–4.

84 Croy, B. A. and Kassouf, S. A. (1989). Evaluation of the murine metrial gland for immunological function. *J. Reprod. Immunol.*, **15**, 51–69.

85 Clark, D. A., *et al.* (1990). Murine pregnancy decidua produces a unique immunosuppressive molecule related to transforming growth factor β-2. *J. Immunol.*, **144**, 3008–14.

86 Lea, R. G., Harley, C. B., and Clark, D. A. (1990). The detection of RNA for TGF-β2 in post-implantation decidual tissue correlates with the detection of a pregnancy associated suppressor factor. *Abstr. Eur. Fed. Immunol. Soc. 10th meeting*, 32–9.

87 Clark, D. A., Slapsys, R. M., Croy, B. A., and Rossant, J. (1983). Suppressor cell activity in uterine decidua correlates with success or failure of murine pregnancies. *J. Immunol.*, **131**, 540–2.

88 Crepau, M. A. and Croy, B. A. (1988). Evidence that specific cellular immunity cannot account for death of Mus caroli embryos transferred to *Mus musculus* with severe combined immune deficiency disease. *Transplantation*, **45**, 1104–10.

89 Kiger, N., Chaouat, G., Kolb, J. P., Wegmann, T. G., and Guenet, J. L. (1985). Immunogenetic studies of spontaneous abortion in mice. Preimmunization of females with allogeneic males. *J. Immunol.*, **134**, 2966–70.

90 Clark, D. A., Chaouat, G., Guenet, J. L., and Kiger, N. (1987). Local active suppression and successful vaccination against spontaneous abortion in CBA/J mice. *J. Reprod. Immunol.*, **10**, 79–85.

91 Gendron, R. L. and Baines, M. G. (1988). Infiltrating decidual natural killer cells are associated with spontaneous abortion in mice. *Cell. Immunol.*, **113**, 261–7.

92 Clark, D. A., Mowbray, J., Underwood, J., and Lidell, H. (1987). Histopathologic alterations in the decidua in human spontaneous abortion: loss of cells with large cytoplasmic granules. *Am. J. Reprod. Immunol. Microbiol.*, **13**, 19–22.

93 Bulmer, J. N. and Johnson, P. M. (1986). The T-lymphocyte population in first-trimester human decidua does not express the interleukin-2 receptor. *Immunology*, **58**, 685–7.

94 Al Jabaari, B., Ladyman, H. M., Larche, M., Sivolapenko, G. B., Epenetos, A. A., and Ritter, M. A. (1989). Elevated expression of the interleukin 4 receptor in carcinoma: a target for immunotherapy. *Br. J. Cancer*, **69**, 910–14.

95 Lewis, C. E., McCarthy, S. P., Richards, P. S., Lorenzen, J., Horak, E., and McGee, J. O. D. (1990). Measurement of cytokine release by human cells. A quantitative analysis at the single cell level using the reverse haemolytic plaque assay. *J. Immunol. Meth.*, **127**, 51–9.

96 Ferry, B. L., Starkey, P. M., Sargent, I. L., Watt, G. M. O., Jackson, M., and Redman, C. W. G. (1990). Cell populations in the human early pregnancy decidua: natural killer activity and response to interleukin-2 of CD56-positive large granular lymphocytes. *Immunology*, **70**, 446–52.

97 Pollard, J. W., Bartocci, A., Arceci, R., Orlofsky, A., Ladner, M. B., and Stanley, E. R. (1987). Apparent role of the macrophage growth factor, CSF-1, in placental development. *Nature*, **330**, 484–6.

98 Bulmer, J. N., Morrison, L., Johnson, P. M., and Meager, A. (1990). Immunohistochemical localization of interferons in human placental tissues in normal, ectopic, and molar pregnancy. *Am. J. Reprod. Immuol.*, **22**, 109–16.

99 Steven, D. H. and Morriss, G. (1975). Development of the foetal membranes. In *Comparative placentation: Essays in structure and function* (D. H. Steven), pp. 58–86. Academic Press, New York.

100 Paterson, W. G. and Grant, K. A. (1975). Advanced intraligamentous pregnancy. Report of a case, review of the literature and a discussion of the biological implications. *Obstet. Gynecol. Survey*, **30**, 715–26.

101 Gambel, P., Croy, B. A., Moore, W. D., Hunziker, R. D., Wegmann, T. G., and Rossant, J. (1985). Characterization of immune effector cells present in early murine decidua. *Cell. Immunol.*, **93**, 303–14.

102 King, A., Birkby, C., and Loke, Y. W. (1989). Early human decidual cells exhibit NK activity against the K562 cell line but not against first trimester trophoblast. *Cell. Immunol.*, **118**, 337–44.

103 Manaseki, S. and Searle, R. F. (1989). Natural killer (NK) cell activity of first trimester human decidua. *Cell. Immuol.*, **121**, 166–73.

104 Ritson, A. and Bulmer, J. N. (1989). Isolation and functional studies of granulated lymphocytes in first trimester human decidua. *Clin. Exp. Immunol.*, **77**, 263–8.

105 Fougerolles, A. R. and Baines, M. G. (1987). Modulation of the natural killer cell activity in pregnant mice alters the spontaneous abortion rate. *J. Reprod. Immunol.*, **11**, 147–53.

106 Ellis, T. M. and Fisher, R. I. (1989). Functional heterogeneity of Leu19 bright+ and Leu19 dim+ lymphokine-activated killer cells. *J. Immunol.*, **142**, 2949–54.

107 Ferry, B. L., Sargent, I. L., Starkey, P. M., and Redman, C. W. G. (1991). Cytotoxic activity, against trophoblast and choriocarcinoma cells, of large granular lymphocytes from human early pregnancy decidua. *Cell. Immunol.*, **132**, 140–9.

108 Chatterjee-Hasrouni, S., Parhar, R., and Lala, P. K. (1984). An evaluation of the maternal natural killer cell population during the course of murine pregnancy. *Cell. Immunol.*, **84**, 264–75.

109 Head, J. R. (1989). Can trophoblast be killed by cytotoxic cells? *In vitro* evidence and *in vivo* possibilities. *Am. J. Reprod. Immunol.*, **20**, 100–5.

110 Zuckerman, F. A. and Head, J. R. (1987). Murine trophoblast resists cell-mediated lysis. I. Resistance to allospecific Cytotoxic T Lymphocytes. *J. Immunol.*, **139**, 2856–64.

111 Zuckermann, F. A. and Head, J. R. (1988). Murine trophoblast resists cell-mediated lysis. II. Resistance to natural cell-mediated cytotoxicity. *Cell. Immunol.*, **116**, 274–86.

112 Drake, B. L. and Head, J. R. (1988). Murine trophoblast cells are susceptible to lymphokine-activated killer (LAK) cell lysis. *Am. J. Reprod. Immunol. Microbiol.*, **16**, 114.

113 Drake, B. L. and Head, J. R. (1989). Murine trophoblast can be killed by allospecific cytotoxic T lymphocytes generated in GIBCO Opti-MEM medium. *J. Reprod. Immunol.*, **15**, 71–7.

114 Hunt, J. S., *et al.* (1989). Products of lipopolysaccharide-activated macrophages (tumor necrosis factor-α, transforming growth factor-β) but not lipopolysaccharide modify DNA synthesis by rat trophoblast cells exhibiting the 80-kDa lipopolysaccharide-binding protein. *J. Immunol.*, **143**, 1606–13.

115 Drake, B. L. and Head, J. R. (1990). Murine trophoblast cells are not killed by tumor necrosis factor-α. *J. Reprod. Immunol.*, **17**, 93–9.

116 Vince, G., Shorter, S. C., Sargent, I. L., Starkey, P. M., and Redman, C. W. G. (1989). The synthesis of TNF by human placental and decidual tissue. *J. Reprod. Immunol.*, (Suppl.), 158.

117 Casey, M. L., Cox, S. M., Beutler, B., Milewich, L., and MacDonald,

P. C. (1989). Cachectin/tumor necrosis factor-α formation in human decidua. Potential role of cytokines in infection-induced preterm labour. *J. Clin. Invest.*, **83**, 430–6.

118 Athanassakis, I. and Wegmann, T. G. (1986). *The immunotrophic interaction between maternal T cells and fetal trophoblast/macrophages during gestation*. In Proceedings of the 3rd International Congress on Reproduction and Immunology (ed. D. A. Clark and B. A. Croy), pp. 99–105. Elsevier, Amsterdam.

119 Wegmann, T. G. (1987). Placental immunotrophism: maternal T cells enhance placental growth and function. *Am. J. Reprod. Immunol. Microbiol.*, **15**, 67–70.

120 Athanassakis, I., Bleackley, R. C., Paetkau, V., Guilbert, L., Barr, P. J., and Wegmann, T. G. (1987). The immunostimulatory effect of T cells and T cell lymphokines on murine foetally derived placental cells. *J. Immunol.*, **138**, 37–44.

121 Rettenmier, C. W., *et al.* (1986). Expression of the human c-fms proto-oncogene (colony-stimulating factor-1 receptor) on peripheral blood mononuclear cells and choriocarcinoma cell lines. *J. Clin. Invest.*, **77**, 1740–6.

122 Garcia-Lloret, M., Guilbert, L., and Morrish, D. W. (1989). Functional expression of CSF-1 receptors on normal human trophoblast. *Placenta*, **10**, 499–500.

123 Arceci, R. J., Shanahan, F., Stanley, E. R., and Pollard, J. W. (1989). Temporal expression and location of colony-stimulating factor 1 (CSF-1) and its receptor in the female reproductive tract are consistent with CSF-1-regulated placental development. *Proc. Natl. Acad. Sci. U.S.A.*, **86**, 8818–22.

124 Uzumaki, H., *et al.* (1989). Identification and characterization of receptors for granulocyte colony-stimulating factor on human placenta and trophoblastic cells. *Proc. Natl. Acad. Sci. U.S.A.*, **86**, 9323–6.

125 Cuturi, M. C., *et al.* (1989). Production of hematopoietic colony-stimulating factors by human natural killer cells. *J. Exp. Med.*, **169**, 569–83.

126 Bartocci, A., Pollard, J. W., and Stanley, E. R. (1986). Regulation of colony-stimulating factor 1 during pregnancy. *J. Exp. Med.*, **164**, 956–61.

127 Ostensen, M. E., Thiele, D. L., and Lipsky, P. E. (1987). Tumor necrosis factor-α enhances cytolytic activity of human natural killer cells. *J. Immunol.*, **138**, 4185–91.

128 Tabibzadeh, S. S., Santhanam, U., Sehgal, P. B., and May, L. T. (1989). Cytokine-induced production of IFN-β2/IL6 by freshly explanted human endometrial stromal cells. Modulation by estradiol-17β. *J. Immunol.*, **142**, 3134–9.

129 Redman, C. W. G. (1989). Hypertension in pregnancy. In *Obstetrics*

(ed. A. Turnbull and G. Chamberlain), pp. 515–41. Churchill Livingstone, Edinburgh.

130 Cordell, J. L., *et al.* (1984). Immunoenzymatic labeling of monoclonal antibodies using immune complexes of alkaline phosphatase and monoclonal anti-alkaline phosphatase (APAAP complexes). *J. Histochem. Cytochem.*, **32**, 219–9.

131 Fleit, H. B., Wright, S. D., and Unkeless, J. C. (1982). Human neutrophil Fc receptor distribution and structure. *Proc. Natl. Acad. Sci. U.S.A.*, **79**, 3275–9.

132 Takeshita, T., Goto, Y., Tada, K., Nagata, K., Asao, H., and Sugamura, K. (1989). Monoclonal antibody defining a molecule possibly identical to the p75 subunit of interleukin 2 receptor. *J. Exp. Med.*, **169**, 1323–32.

133 Larche, M., *et al.* (1988). Functional evidence for a monoclonal antibody that binds to the human IL-4 receptor. *Immuology,* **65**, 617–22.

134 Starkey, P. M. (1991). Expression on cells of early human pregnancy decidua of the p75, IL-2 and p145, IL-4 receptor proteins. *Immunology*, **73**, 64–70.

135 King, A. and Loke, Y. W. (1990). Human trophoblast and JEG choriocarcinoma cells are sensitive to lysis by IL-2 stimulated decidual NK cells. *Proceedings of the Eur. Fed. Immunol. Soc. 10th meeting, Edinburgh*, U.K.

136 Shorter, S. C., Vince, G. S., and Starkey, P. M. Production of granulocyte-colony stimulating factor at the materno-foetal interface in human pregnancy. *Immunology*. (In press.)

Index